Modern Spectrum Analyzer Theory and Applications

Modern Spectrum Analyzer Theory and Applications

Morris Engelson

Copyright © 1984
ARTECH HOUSE, INC.
610 Washington Street, Dedham, Massachusetts

Printed and bound in the United States of America

All rights reserved. No parts of this book may be reproduced or utilized in any form or by any means, electronic or mechanical, including photocopying, recording, or by an information storage and retrieval system, without permission in writing from the publisher.

Standard Book Number: 0-89006-150-5

Library of Congress Catalog Number: 84-045200

To Jan
You made it possible.

CONTENTS

Preface — 1984 Edition xi
Preface — 1974 Edition xii
Acknowledgment xiii

Chapter 1   **Spectrum Analyzers**
               1.1.   Introduction 1
               1.2.   Constructing a Spectrum Analyzer 2
               1.3.   The Sweeping Signal System 5
               1.4.   Types of Signals 6
               1.5.   Examples 8
               1.6.   Exercises 10

Chapter 2   **Spectrum Theory**
               2.1.   Time and Frequency Domain 11
               2.2.   Orthogonal Functions 12
               2.3.   The Properties of Sinewaves 16
               2.4.   The Fourier Series 22
               2.5.   Do Spectral Lines Exist? 24
               2.6.   Response of Circuits to Signals 25
               2.7.   Proofs and Examples 27
                       2.7.1.   Orthogonal Functions 27
                       2.7.2.   Complex Notation 28
                       2.7.3.   Proof that the Truncated Fourier Series
                                Provides a Least Squared Error Fit 28
               2.8.   Exercises 31

Chapter 3   **Fourier Analysis**
               3.1.   Introduction 33
               3.2.   Fourier Series................................ 33
               3.3.   Fourier Applications 35
               3.4.   Superposition 39
               3.5.   Gibbs Phenomenon 42
               3.6.   Continuous-Dense Spectrum 44

	3.7.	Fourier Integral 49
	3.8.	Reconciling Theory and Measurement 50
	3.9.	Proofs and Tables 52
		3.9.1. The Integral Equations for the Fourier Coefficients 52
		3.9.2. The Impulse Function 54
		3.9.3. Properties of Fourier Transforms......... 56
	3.10.	Examples 57
		3.10.1. Using One Spectrum to Derive Another 57
		3.10.2. The Sine Integral, Si(x) 62
		3.10.3. Rectangular Pulse Analysis 63
	3.11.	Exercises 65
Chapter 4	**Modulation Theory**	
	4.1.	Introduction 69
	4.2.	Amplitude Modulation......................... 70
	4.3.	Angle Modulation 73
	4.4.	Bessel Functions 76
	4.5.	The FM Spectrum 78
	4.6.	Combined Modulation 81
	4.7.	Time Domain and Frequency Domain 83
	4.8.	Examples 86
	4.9.	Exercises 87
Chapter 5	**The Sweeping Signal Spectrum Analyzer**	
	5.1.	Introduction 89
	5.2.	The CW Response 90
	5.3.	Pulsed Signals 95
	5.4.	Sensitivity................................... 99
	5.5.	Convolution 102
	5.6.	Examples 103
	5.7.	Exercises 105
Chapter 6	**The Measurement Problem**	
	6.1.	Introduction 107
	6.2.	Types of Measurements 107
	6.3.	Measurement Limitations 108
	6.4.	Spurious Response Basics 108
		6.4.1. Swept IF 108
		6.4.2. Swept Front-End 111
	6.5.	More on True Signal Identification 113
	6.6.	Other Spurious Responses 115
	6.7.	Exercises 117

Chapter 7	**Amplitude Modulation**	
	7.1.	Basic Relationships 121
	7.2.	Measuring AM 122
	7.3.	Other Forms of AM 123
	7.4.	Exercises 126

Chapter 8	**Frequency Modulation**	
	8.1.	Basic Relationships 129
	8.2.	Narrowband FM 131
	8.3.	Wideband FM............................. 135
		8.3.1. Example 1 136
		8.3.2. Example 2 138
		8.3.3. Example 3 138
	8.4.	Ultrawideband FM 141
	8.5.	Determining Modulation Rate for Unresolved Signal 142
	8.6.	Combined AM and FM 143
	8.7.	Multitone FM............................. 145
	8.8.	Intensification Effects 146
	8.9.	Exercises 147

Chapter 9	**Pulses**	
	9.1.	Mathematical Relationships 149
	9.2.	Making the Measurement 151
	9.3.	Radar Performance 154
	9.4.	Effect of Pulse Shape...................... 155
	9.5.	Effect of FM 157
	9.6.	Percentage of Missing Pulses Determination...... 160
	9.7.	Measuring Modulator On/Off Ratio 161
	9.8.	Effect of Control Settings 164
		9.8.1. Repetition Rate 164
		9.8.2. Dense Versus Line Spectrum 164
		9.8.3. Fine Detail 166
		9.8.4. Sensitivity and Dynamic Range 166
		9.8.5. Display Intensification Effects 168
	9.9.	Reconciling Theory and Measurement 171
	9.10.	Determining Impulse Bandwidth 173
	9.11.	Exercises 177

Chapter 10	**Miscellaneous Applications**	
	10.1.	Waveform Analysis 179
		10.1.1. Squarewave 179
		10.1.2. Symmetry of Adjustment 181

	10.1.3.	Squarewave on Carrier 182
10.2.		Random Noise Measurement 183
	10.2.1.	Relative Measurement 183
	10.2.2.	Absolute Measurement................ 185
	10.2.3.	Making the Measurement 187
	10.2.4.	Examples 189
10.3.		Distortion Measurement 192
10.4.		Component Transfer – Characteristic Measurement. 193
10.5.		Synchronized Sweeper Techniques 194
10.6.		EMI Measurements 197
10.7.		Telemetry Subcarrier Tests 198
10.8.		Doppler Velocity Measurement 200
10.9.		Using Transducers 201
10.10.		Sensitivity Improvement 203
10.11.		Digital Radio Signals 204
10.12.		Exercises 206

Chapter 11   **Specifications**
          11.1.   Introduction 209
          11.2.   Sensitivity 209
          11.3.   Resolution 210
          11.4.   Stability................................. 214
          11.5.   Display Flatness/Frequency Response 219
          11.6.   Dynamic Range 220
          11.7.   Combining Specifications 222

Appendix
dB, dBm 225
Bessel Functions 228
Fourier Analysis 231
CW Sensitivity 231
Resolution Bandwidth 233
Pulse RF 234
Random Noise 237
Intermodulation Dynamic Range 238
Symbols 238

Bibliography 243

Index 251

PREFACE – 1984 EDITION

Many changes in instrumentation and applications have occurred since *Spectrum Analyzer Theory and Applications* was published in 1974. The present volume, *Modern Spectrum Analyzer Theory and Applications,* is a revision and update of that work. Fourier Theory, or the basic theory behind the sweeping spectrum analyzer developed during WWII, is, of course, not affected. The major changes will be found in the later, practical chapters. Here the reader will find an updated discussion on pulsed signals, a thorough introduction to random noise measurements, a comparison of various kinds of spectrum analyzers including the newest computer enhanced variety, among many other topics, as well as an updated bibliography to take into account developments over the last 10 years. Symbols and nomenclature has been updated in accordance with the latest international standards. The aim of this book remains unchanged, namely, it is a reference for practicing engineers and can be used as a text in a course on the frequency domain.

PREFACE – 1974 EDITION

This book is primarily concerned with the problems of measurements in the frequency domain by means of Spectrum Analyzers. Thus circuit design or construction details are not considered. Basic system parameters are, however, discussed in some detail since these have a direct bearing on the interpretation of measurement data. Two types of signals are treated in detail: those composed of discrete or line spectra and those composed of continuous or dense spectra. Continuous wave (CW) or sinusoidal amplitude modulation (AM) is an example of the former, while pulsed-RF is treated as the latter. The third class of signals comprising random variables and requiring statistical methods are excluded from the detailed discussion, though some applications are included.

The discussion follows a dual approach. Chapters one through five have a mathematical, process-oriented approach; the latter part of the volume applies the former theory to specific measurement problems. Thus, those wishing to avoid mathematical complexity might go directly to chapter six. Those who wish a somewhat deeper understanding should read selectively in the first five chapters since the more abstruse material has been relegated to the proofs at the end of each chapter.

A casual reader might find the material or text uneven. An elementary discussion of the properties of sinewaves does not seem to fit into a chapter which is full of integrals and partial differentials. The reason for the above is that the book addresses itself to three related, yet different, problems. First, an attempt is made to present a unified mathematical, physical and philosophical rationale for the use of spectrum analyzers. While most of this material has been available previously, it was never present in one source and with a consistent point of view. Another application of this book is as a reference volume for users of spectrum analyzers. Here, chapters six through ten will have the greatest utility. However, those who wish to use spectrum analyzers in new and as yet untried areas should find the proofs and references of interest. Finally, this volume can be used as a text for a course in the frequency domain. To this end, the book includes many detailed examples as well as homework exercises.

An attempt has been made to help the reader find his way around. Specialized terms are discussed in chapter eleven. The appendix includes an extensive list and definition of symbols as well as other useful formulas, graphs and tables. The volume is concluded with a bibliography on spectrum analysis and related topics.

ACKNOWLEDGMENT

The author wishes to thank Tektronix, Inc. for unrestricted use of laboratory facilities and permission to reprint material previously copyrighted by Tektronix, Inc.

Chapter 1
Spectrum Analyzers

1.1. INTRODUCTION

All electronic signals, indeed all natural phenomena, can be described either as a function of time or frequency. When a phenomenon is cyclical, having a definite periodicity, the basic relationship between frequency and time interval is fairly simple, one being essentially the inverse of the other. In the case of random phenomena, one has to use statistical methods, but the concept of the duality of time and frequency is still useful. The concept of frequency as considered here presupposes time duration* — where time is a basic property of the universe we live in and frequency is related to time through the cyclical or periodic nature of the phenomena under discussion.

Just as the oscilloscope's basic function is to display the time characteristics of phenomena, the spectrum analyzer's function is to display the frequency characteristics of phenomena. It should be recognized that the two descriptions of the same phenomena — time domain for the oscilloscope and frequency domain for the spectrum analyzer — are not independent. If one of the two is known, the appropriate mathematical rules or equations lead to the other. The question of which description is the more basic, time domain or frequency domain, is difficult to answer. It can be argued that time is the basic natural phenomenon and that the frequency concept is derived from it since a universe without time duration is inconceivable. It can, however, also be argued that it is the periodicity of natural phenomena that makes for "time." Certainly in modern physics, such as relativity theory, it is considered that time stems from the existence of matter, and hence movement and periodicity, and that time without matter (Newton's absolute time) has no meaning.

For discussion purposes, time will be the basic concept and frequency will be derived from it. This is convenient because oscilloscopes are constructed to display an enhanced (amplified and/or sampled) version of the incoming signal, whereas spectrum analyzers are constructed to obtain the frequency-domain

---

*One could consider frequency in more general terms — i.e., one could say that a topographical distribution of hills is cyclical with x hills per mile. Time need not enter such a discussion.

characteristics of the incoming signal by computations or other operations performed on a time-domain input. Thus, while an oscilloscope is generally recognizable as such from its block diagram, a spectrum analyzer may be difficult to recognize, since the computation function or analog operation can be performed by diverse means.

Some of the methods that can be used to obtain a frequency-domain presentation are now described.

1.2. CONSTRUCTING A SPECTRUM ANALYZER

A computer can be used to perform the appropriate computation for going from the time to the frequency domain. With the advent of the time-saving Fast Fourier Transform technique, this method of spectrum analysis can be quite attractive. Nevertheless, the computer programmed as a spectrum analyzer is still a rarity, which is used primarily for low frequency applications.

Besides the computer method, the frequency composition of a complex signal can also be obtained by separating the several components in a contiguous bank of filters. Such a system of contiguous-filter passbands can be constructed from real physically existing filters. Figures 1-1(A) and 1-1(B) illustrate the block diagram and frequency characteristics of such a system. The composite time-domain signal is fed to the multicoupler, which distributes it equally among the several filters. Each filter will respond only to inputs within its passband. Hence, by observing the amplitudes of the outputs of the various filters one can determine which frequencies and what amplitude levels are present in the composite input signal.

The narrower the filter bandwidth, B, the better our ability to determine the precise frequencies of the signal components. This discrimination between signals having closely spaced frequencies is called resolution — the narrower B, the better the resolution. An analogous example is the resolution of a microscope, where an improvement in resolution refers to an increased ability to separate visually several small particles. The range of frequencies that can be analyzed is the total frequency band covered by the set of contiguous filters. If there are a total of n filters each having a bandwidth B, then the total frequency range is simply the product nB. This used to be known as the dispersion, a word borrowed from optics. The current term is frequency span, or span. As the filter bandwidth is decreased in order to improve resolution, it becomes necessary to increase the number of filters (n) correspondingly if the span is not to decrease. Thus, the number of filters and indicators in such a system can get very large.

One way to reduce the cost of such a system is to use one recorder, or indicator, which is commutated between the various filters, as is depicted in Figure 1-2. Here, instead of determining frequency by seeing which indicator shows an out-

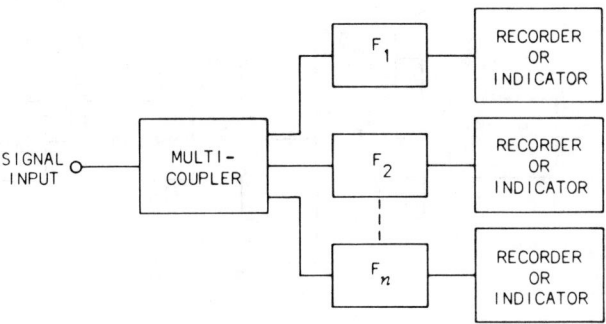

(A) Block Diagram

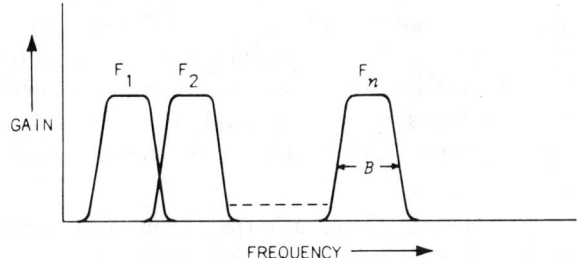

(B) Frequency Characteristics

Figure 1-1 Contiguous Filter Bank

put, we determine frequency by correlating the time of the output with the sequence of the commutator. For example, if the commutator is connected to each filter for half a second and it takes a half-second travel time to go from filter to filter, an output at 10 to 10.5 seconds after start means a signal corresponding to the frequency of the eleventh filter. With the appropriate recorder speed, for example one inch per second, we can transform the recorder time axis into a frequency axis. Thus, for example, a signal indication positioned 10 inches from the start means a signal at the frequency of the eleventh filter.

Superficially, the systems shown in Figures 1-1(A) and 1-2 seem to accomplish the same thing. There is, however, one major difference. Whereas the first system will, at least in theory, show all signals no matter how short their duration, the second system is limited in this respect by the speed of the commutator and the time constants or memory associated with the system. Thus, the convenience

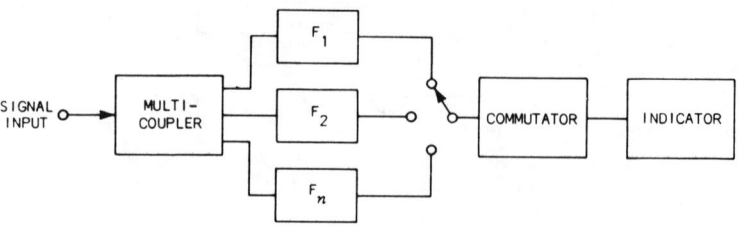

Figure 1-2 Filter Bank System Using Commutator

of reduction in system complexity is paid for by the loss of some capability.

The system shown in Figure 1-2 is still quite cumbersome; it may require hundreds of filters to obtain the desired resolution and span. However, since filters no longer are utilized on a continuous basis, if a single filter whose center frequency is switched or tuned in lieu of the commutator, the system depicted in Figure 1-3 results. The characteristics of such a system are best understood by considering the relationships shown in Figure 1-4, where a filter having a bandwidth B is assumed to be tuned over the frequency range f_1 to f_3 during the time interval T.

During the same time interval, the indicator (paper chart recorder or CRT) changes position from 0 to 10 divisions. For example, zero position corresponds to f_1, 10 divisions corresponds to f_3 and 5 divisions corresponds to the center frequency between f_1 and f_3, namely $(f_1 + f_3)/2$. In addition, Figure 1-4 also shows a signal at frequency f_2. The signal is shown as a straight line in the time-frequency diagram, meaning that it has constant frequency as a function of time. The effect of the tuning or sweeping filter intercepting the signal is indicated by the pulses on the horizontal position scale. The width of the pulses traced by the indicator is $\tau = BT/(f_3 - f_1)$, namely the time that the signal frequency is within the passband of the filter.

This system is relatively simple and compact, but there are practical difficulties. Problems stem from the present state of the art in electronically tuneable-filter construction, which have much wider bandwidths than is desired for most applications. Thus, spectrum analyzers of this type have somewhat limited utility.

The transformation from time domain to frequency domain is accomplished by the relative translation or movement in frequency between the filter and the signal in the system of Figure 1-4. It does not matter whether it is the filter or the signal frequency that is changing or translating. Thus, the same end result as that of Figure 1-4 should be obtained by using a stationary filter and a translating or, to use the more common name, sweeping signal. The time/frequency-position relationships for such a system, the superheterodyne system, are shown in Figure 1-5. Here the filter is shown as having a passband (B) and a constant unchanging center frequency (f_2). When there is a signal whose frequency falls within the passband of the filter, there is an output. The result is the pulse on the position scale whose position corresponds to a frequency, f_2, and where the pulse width $\tau = BT \big/ (f_3 - f_1)$ is a measure of system resolution.

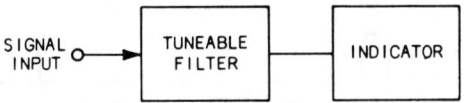

Figure 1-3 Tuned-Filter Spectrum Analyzer

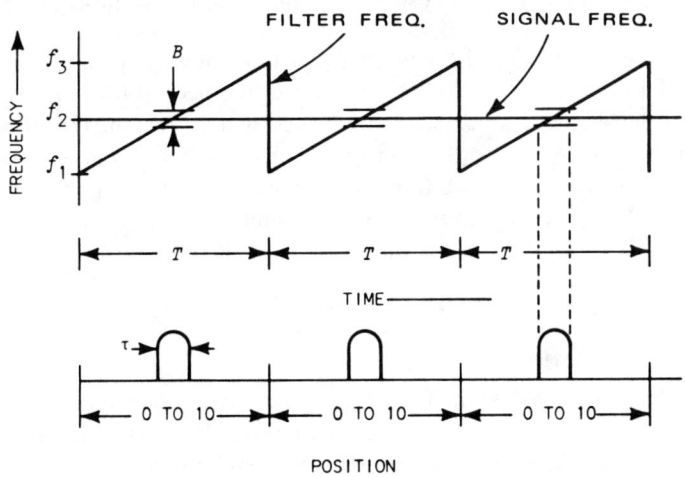

Figure 1-4 Frequency/Time-Position Relationships for Tuned-Filter System

1.3. THE SWEEPING SIGNAL SYSTEM*

The superheterodyne or sweeping-signal system is based on the use of a mixer, which, for the present, can be considered an idealized three-port black box. The three ports provide for a signal input, local oscillator input, and intermediate frequency (IF) output. The idealized mixer produces an IF output that has the

---

*Detailed characteristics are described in Chapter 5.

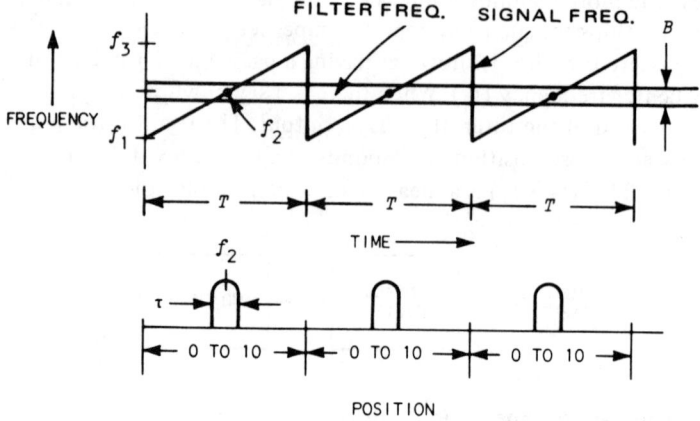

Figure 1-5 Frequency/Time-Position Relationships for Swept-Signal System

amplitude characteristics of the input signal and whose frequency characteristics consist of an algebraic combination (sum or difference) of the LO frequency and the input signal frequencies. What amounts to signal sweeping is produced by sweeping the local-oscillator frequency, which produces a swept IF output. Figure 1-6 is a block diagram and frequency-time display showing the effect of multiple signal frequencies; for each signal frequency generated, a separate IF frequency sawtooth is produced through one of the relationships:

$$f_{IF} = \begin{array}{l} f_{LO} + f_{RF} \\ f_{LO} - f_{RF} \\ f_{RF} - f_{LO} \end{array}$$

Figure 1-6(B) is drawn for the $f_{IF} = f_{LO} - f_{RF}$ relationship. As the IF frequency sawtooth passes through the filter passband of bandwidth B and center frequency f_0, a pulse of width τ is generated. One such pulse is generated for every signal frequency present, with pulse height being proportional to signal and amplitude. In Figure 1-6(B) it is assumed that the amplitude of the signal at f_2 is larger than that at f_1.

1.4. TYPES OF SIGNALS

The examples considered thus far are based on the assumption that the input can be considered as consisting of several sinewaves and that the response of a system to a continuous wave (CW) signal is all that's needed since the rest follows from simple superposition. But this assumption is not always warranted. A case in point is pulsed RF. The output pulse shown in Figure 1-6(B) cannot be obtained if the input-pulse duration is less than the width τ and the response

Spectrum Analyzers

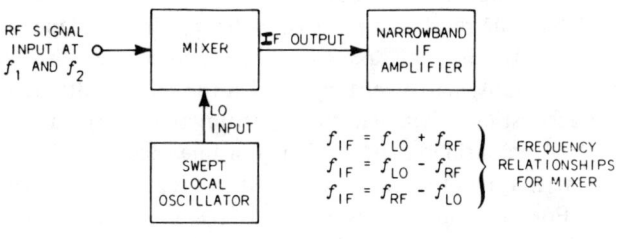

(A) Basic System

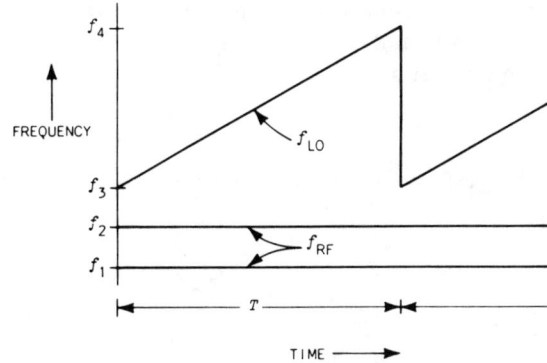

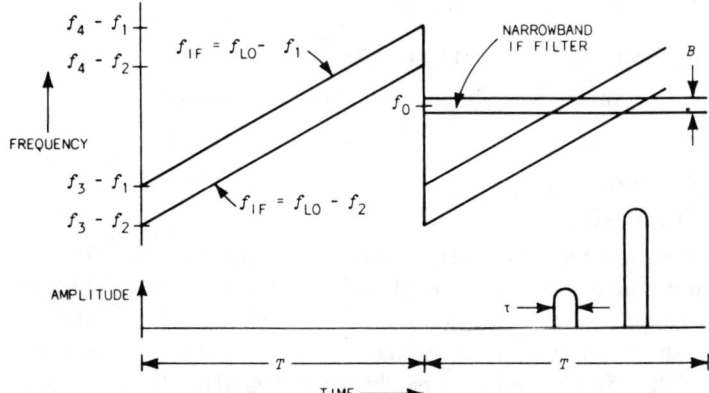

(B) Frequency-Time Relationships Using $f_{IF} = f_{LO} - f_{RF}$ Mixing System

Figure 1-6 Superheterodyne System, Responding to Two Input Signals of Different Frequencies and Amplitude

of the system to pulsed inputs is much more complicated than that for CW-type inputs. In the CW case, the result is a pulse of width τ, which is dependent on the resolution bandwidth, the span, and the sweep time; in the pulsed RF case, there is a pulse considerably narrower than τ, with the pulse width determined by the signal characteristics rather than the spectrum-analyzer parameters. The former type of signal is considered as consisting of line spectra – discrete CW components; the latter type of signal is described by a continuous spectrum or a dense spectrum. For the moment it is helpful to recognize one basic difference: In the CW case, the final result is a pulse tracing out the shape of the resolution filter, which can be considered as the steady-state response of the narrowband IF amplifier; in the continuous-spectrum case, the narrowband amplifier has to respond to a fairly narrow pulse resulting in a transient rather than steady-state response. The significance of this and other differences between these two classes of signals is considered in Chapter 5.

1.5. EXAMPLES

Referring to Figure 1-6(A), assume that the following numbers apply:

1. The narrowband IF
   $f_0 = 75$ MHz
   $B = 100$ kHz
2. The swept local oscillator
   $f_3 = 270$ MHz
   $f_4 = 280$ MHz
   Sweeptime $T = 10$ ms

Consider the response to a CW signal at 200 MHz.

The signal can combine with the local oscillator as follows:

$$f_{IF} = \begin{array}{l} f_{LO} + f_{RF} \\ f_{LO} - f_{RF} \\ f_{RF} - f_{LO} \end{array}$$

In our case, we are only interested in the results where $f_{IF} = f_0 = 75$ MHz. This means that we are interested in the relationship $f_{LO} - f_{RF} = f_{IF}$, since $275 - 200 = 75$. This happens when the local oscillator is 275 MHz, which is the center of its sweep, 5 ms from sweep start. Actually we get an output not only at 75 MHz but at 75 MHz $\pm$ 50 kHz since the bandwidth is 100 kHz. Hence the result is a pulse of width $\tau = (B/S)T = [0.1/(280 - 270)] 10 = 0.1$ ms centered at 5 ms from the start of the sweep. Assuming that this pulse were displayed on a CRT having 10 horizontal divisions at 1 ms/DIV (same as the sweeping LO), we would have a pulse occupying 0.1 divisions. It is interesting to note that the pulse continues to occupy 0.1 divisions regardless of what the sweeptime T is, so long as the sweeping oscillator and the time base of the CRT are identical. Thus, if T is made 100 ms, τ becomes 1 ms and at 10 ms/DIV still occupies 0.1 divisions.

Consider now that the signal consists of two components — one at 200 MHz and the other at 202 MHz. The result is two output pulses, one at 5 ms from the start and the second at 7 ms from the start. These would appear on the CRT at 5 and 7 divisions respectively. The relative amplitudes of these pulses would be in the same proportion as the relative amplitudes of the signal components. Proceeding in similar fashion, we see that the horizontal CRT scale can be considered as a frequency scale where the left-hand edge represents 195 MHz and the right-hand edge represents 205 MHz.

Consider now that the signal has a third component at 350 MHz. This too will appear on the CRT via the relationship $f_{RF} - f_{LO} = f_{IF} = 350 - 275 = 75$. Yet the CRT has not been calibrated for it, since we've assumed that our frequency base is 195-205 MHz. Such a signal, which does not conform to the frequency calibration of the CRT, is called a spurious response. There are many types of spurious responses. This particular spurious response is called the image. Let us now go back to our original signal at 200 MHz and see what happens to it as a function of the spectrum analyzer control settings. As previously determined, this signal appears on the CRT as a response 0.1 DIV wide located in the center of the CRT with an amplitude which is proportional to the input level.

1. Changing the sweeptime: As previously determined, this should have no effect on the appearance of the pulse. Thus, the width of the signal pulse is a true measure of relative resolution since it is, at least in theory, dependent only on resolution and span. In actuality, if the sweeptime is reduced too much, there will be major changes in what appears on the CRT. This aspect will be discussed in the section on spectrum analyzer limitations.

2. Changing the local-oscillator center frequency: Let the sweeping local oscillator operate from 271-281 MHz. Thus, 275 MHz occurs four-tenths from the beginning and our signal pulse will move from the fifth to the fourth graticule line on the CRT. This is because the CRT frequency base has now changed from 195-205 MHz to 196-206 MHz.

3. Changing the sweeping-oscillator sweep width or span: Let the sweeping oscillator operate from 272.5-277.5 MHz for a total excursion of 5 MHz. The pulse width is $\tau = (B/S)T = (0.1/5)T$ seconds which occupies a physical distance of $(0.1/5)T$ seconds $\times$ (10 CRT divisions/T) = 0.2 DIV, or twice the previous width. The frequency base of the CRT has likewise been changed from 195-205 MHz to 197.5-202.5 MHz.

4. Changing the resolution bandwidth: Let the resolution bandwidth B = 50 kHz. The only effect is to reduce the signal pulse width to $\tau = (0.05/10)\,\mathrm{T}$ seconds or 0.05 divisions wide on screen.

1.6. EXERCISES

1.1. Given: A sweeping spectrum analyzer has the following parameters: The sweeping local oscillator sweeps from 250 kHz to 350 kHz; the input signal frequency range is dc to 100 kHz; and the resolution bandwidth is 1 kHz.
   a. What is the center frequency of the IF amplifier (f_{IF})?
   b. What input frequency will result in a response at the center of the CRT indicator?
   c. What input frequency outside the specified dc - 100 kHz range can cause a spurious response at the center of the CRT indicator?
   d. Can this spectrum analyzer distinguish between two signals whose frequencies differ by 500 Hz? by 5 kHz? by 50 kHz?
   e. Observing the IF output with a separate oscilloscope, what will be observed for a single frequency CW input?
   f. If the spectrum analyzer sweep time is 1 s/DIV (10 s total), what is the width of the output pulse (τ)?
   g. What are the necessary characteristics of a filter, which when positioned in front of the input mixer, will prevent spurious responses such as in question c without affecting the desired frequency range?

Chapter 2
Spectrum Theory

2.1. TIME AND FREQUENCY DOMAIN

Any motion, or to use the electronic term waveform, which repeats itself as a function of time, is called cyclical.

When the waveform repeats in equal intervals of time, it is considered periodic.

When a waveform is generated by the retracing of the same path, such as the back-and-forth motion of a pendulum or the back-and-forth transfer of charge in an LC circuit, is called oscillatory.

One complete oscillation means that a round trip is completed, e.g., from A to B and back to A again.

The period (T) of the oscillation is the time required to complete one oscillation.

The frequency (f) is the number of oscillations per unit time, i.e., $f = 1/T$, $T = 1/f$.

Consider a periodic waveform such as the squarewave shown in Figure 2-1(A). This waveform can be described as having a period T or a frequency $f = 1/T$; either of these statements combined with the statement that the amplitude is A gives a complete description of the squarewave. However, when comparing this description with the graphical representation, it is observed that the description in terms of the period T is easier to use since the description in terms of frequency is not applicable to the coordinate system of the graph without mathematical computation. Yet it is often more meaningful to describe a phenomenon in terms of frequency rather than time duration. This leads to the desirability of producing a graphical representation of the squarewave as a function of frequency rather than time. Such a graphical representation is shown in Figure 2-1(B). This figure, however, only conveys the appropriate information to those who know the conventions used. Thus, the basic waveform must be specified as a squarewave, not a triangle, trapezoid, or sinusoid.

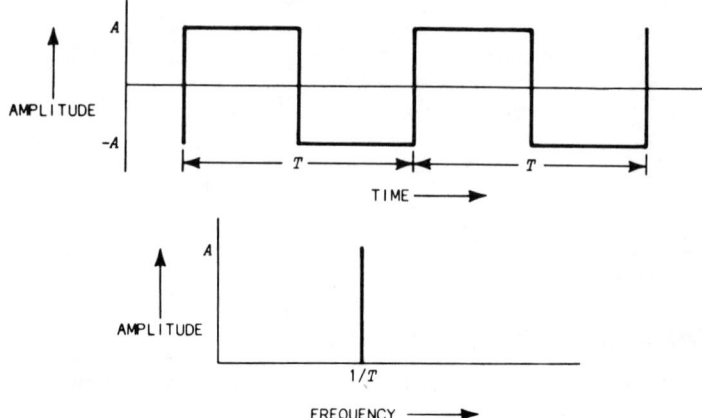

Figure 2-1 Periodic Waveform
    (A) Time Domain
    (B) Frequency Domain

Knowing the conventions, it is possible to represent all kinds of waveforms in the frequency domain by breaking them up into squarewaves. Such a representation is shown in Figure 2-2. Here the complex pulse shape of Figure 2-2(C) is represented as the sum of the two squarewaves shown in Figures 2-2(A) and 2-2(B). These squarewaves are in turn represented in the frequency domain by Figure 2-2(D). As far as information content is concerned, Figures 2-2(C) and 2-2(D) are completely equivalent, each being derivable from the other.

It should be noted that there is nothing sacred about using squarewaves as the basic waveform; triangles, trapezoids, or sinusoids might be much more convenient.

2.2. ORTHOGONAL FUNCTIONS

The most important requirement of the basic waveform is that as many as possible (preferably all) other waveforms should be disassociable into a combination of basic waveforms. Though many types of waveforms, or to use the proper mathematical terminology functions, can be used as the basic function, it can be shown* that sets of functions that possess the property of orthogonality fulfill the above requirement best. This is because orthogonal functions have the least degree of correlation with respect to each other and will generally convey the most information for the least number of terms.

---

*See, for example, Whittaker & Watson's *Modern Analysis* for a detailed mathematical exposition on the expansion of functions in infinite series using both orthogonal and nonorthogonal functions.

Spectrum Theory

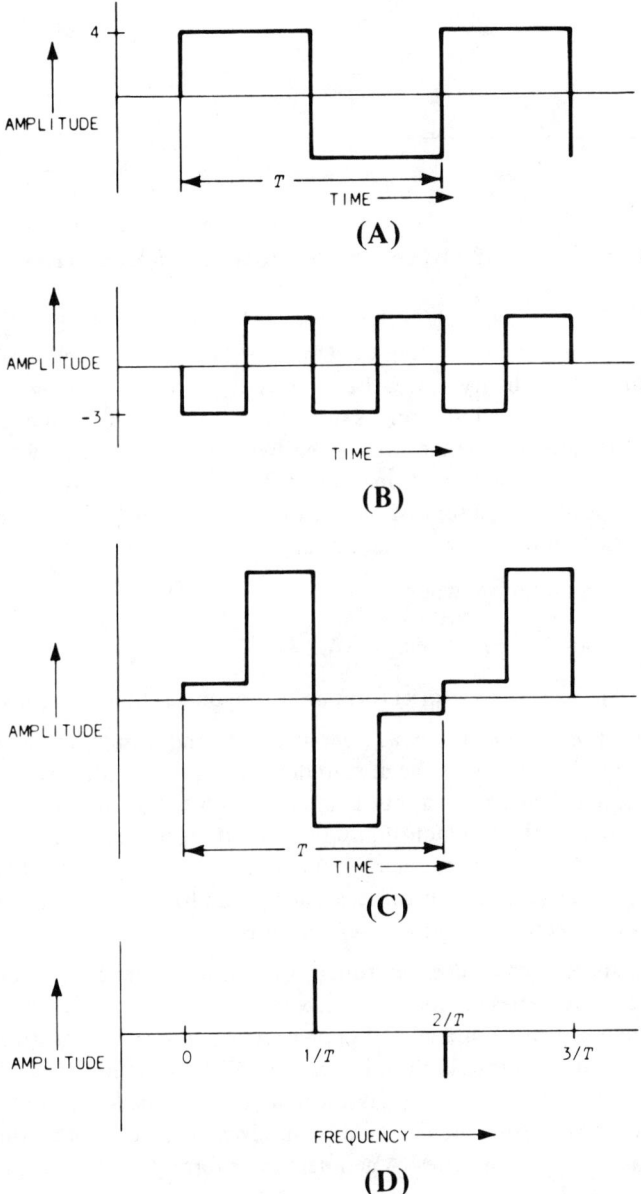

Figure 2-2 Time and Frequency Domain Representation Using Squarewave as Basic Frequency Function

Figure 2-3 Projection of ℓ When α ≠ 90°; Projection f(X) is Zero When α = 90°

The word orthogonal comes from the words orthos meaning right and gonia meaning angle. In ordinary usage, the word is defined as pertaining to right angles. The mathematical meaning is more precise but based on the same classical roots. It is based on the fact that when two lines or planes are at right angles to each other, the projection of one onto the other is zero, as is illustrated in Figure 2-3. Specifically, a set of functions is orthogonal when the integral between specified limits of the product of any two functions is zero.

In mathematical notation, when

$$\int_a^b f_m(x)f_n(x)dx = 0, \text{ when } m \neq n,$$

then $f_1(x), f_2(x) \ldots f_m(x), f_n(x)$ form an orthogonal set of functions.

One can interpret orthogonality as a geometrical condition by considering that the result of an integration is the area under the curve bounded by the function being integrated. Thus, when a set of functions is orthogonal, the area under the curve generated by the product of any two functions, except the function times itself, is zero. A physical example of orthogonal functions is a set of three mutually perpendicular vectors. For this case, according to the equation, the projection of any one vector upon any other vector is always zero.

There are many sets of orthogonal functions. For example: Sinewaves, Bessel functions, and the series composed of $1, x, x^2 - (1/3), x^3 - (3/5) x, \ldots$ taken between the limits of ±1, are all orthogonal. An analytical demonstration of the orthogonality of this series is given in Section 2.7.1. While more obvious graphical examples may be found using discontinuous functions, such as Walsh functions (square waves), they do not lend themselves to the ease of mathematical analysis exhibited by this series. A geometrical interpretation of this series is shown in Figure 2-4.

Spectrum Theory

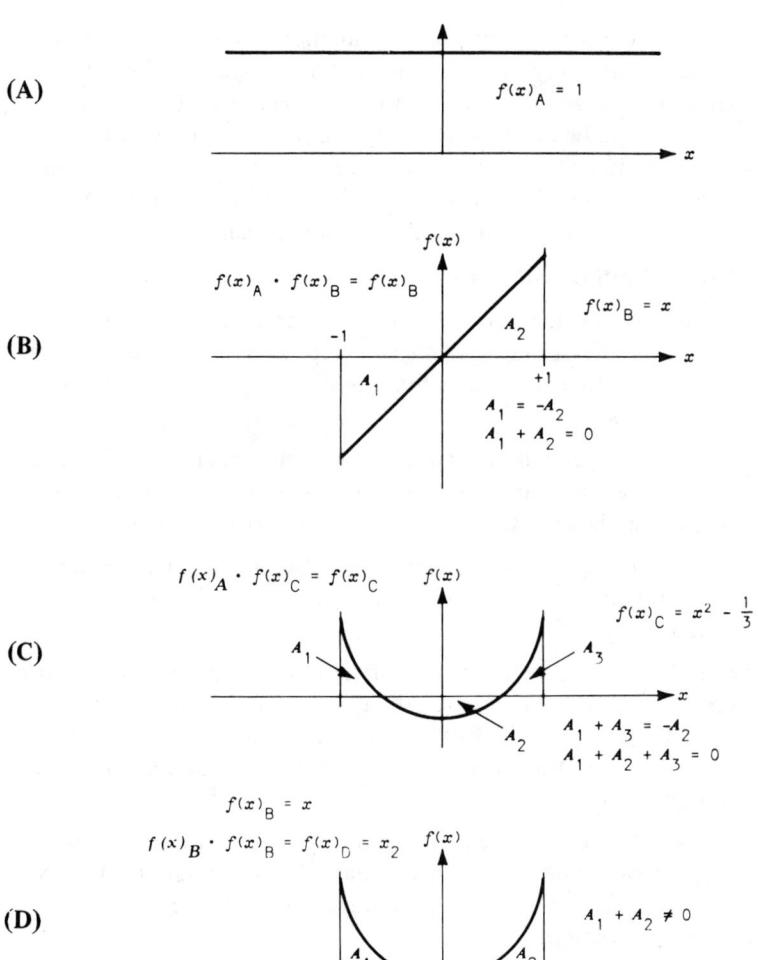

Figure 2-4 Geometrical Interpretation of the Orthogonal Series; $1, X, X^2 - 1/3$, ... The Integral Between The Limits of the Product of Any Two Terms is Zero as Demonstrated in (A), (B), and (C), While the Integral of the Product of Any Term Itself Is Not Zero, as Illustrated in (D)

As previously indicated, a series of orthogonal functions serves our purposes best. This is because almost any function* defined over a specific interval, such as $+\pi$ to $-\pi$, can be expanded in a set of orthogonal functions. Thus, from the point of view of what can be done theoretically, Bessel functions, which we shall discuss in connection with frequency modulation, are just as good as sines and cosines. The question of which orthogonal set of functions to use must, therefore, be settled on the merits of practicality rather than theory.

2.3. THE PROPERTIES OF SINEWAVES

Among the various sets of orthogonal functions, that consisting of sines and cosines comes closest to describing the behavior of physical systems and is one of the easiest to manipulate mathematically. The sine functions have the following important properties:

1. The sine function is generated in connection with motion around a circle, as will be demonstrated later. Since much machinery is based on circular motion, the sine function can be used to describe physically existing situations.

2. Most physical processes that are undulatory are also periodic. Unlike the orthogonal Bessel functions, for example, which are undulatory but not with a constant period, the sine functions are periodic.

3. Many diverse phenomena, such as the oscillation of a weight on a spring, the swing of a pendulum, or the oscillation of current in an LC circuit, are basically sinusoidal. In particular, the fact that the natural behavior of electrical circuits is sinusoidal is of the greatest importance in the choice of sinewaves as our basic waveform.**

4. Sinewaves possess the remarkable mathematical property: The basic description remains invariant under various mathematical transformations. For example, except for a change in phase, a sinewave remains a sinewave with integration or differentiation.

Based on reasoning such as above, the conclusion is reached that, among the

---

*There are some unimportant exceptions.

**Actually, the damped rather than the constant-amplitude sinusoidal oscillation is the natural response of real networks, since all physical networks contain some losses. However, the undamped sinewave is so much easier to manipulate that it has become the universal choice as the basic waveform.

Spectrum Theory

various sets of orthogonal functions, sinewaves are both easier to manipulate and come closer to describing physical processes than any other type of function. Hence, when describing complex waveforms by breaking these up into a sum of more elementary waveforms, the sinewave is chosen as the basic waveform.

Now consider some of the terminology and properties associated with sinewaves.

The behavior of many physical systems is described by a differential equation of the form $d^2y/dt^2 + \omega^2 y = 0$, a solution of which is $y = A \cos \omega t + B \sin \omega t$. In electronics such an equation is basic to LC circuits

$$\frac{d^2Q}{dt^2} + \left(\frac{1}{\sqrt{LC}}\right)^2 Q = 0$$

where L, C, and Q are inductance, capacitance, and charge respectively. The solution is in the form of a sinusoidal oscillation at angular frequency $\omega = 1/\sqrt{LC}$.

The functions $y = A \cos \omega t$ or $y = B \sin \omega t$ are often referred to as circular trigonometric or just circular functions. These functions are also sometimes connected with the words simple harmonic motion, since they describe the simple harmonic motion of a point around a circle, as well as other physical phenomena. The words sinewave or sinusoidal will be used whenever possible. It should be understood that the general word sinewave refers to both $\sin \theta$ and $\cos \theta$.

The general expression for a sinewave is $y = A \sin \theta$. A represents the amplitude, while θ represents the angle. In electrical problems, θ is usually replaced by the time-dependent expression $\omega t + \alpha$, where ω is the radian frequency or angular velocity, the combined quantity $(\omega t + \alpha)$ is the phase, while the fixed angle α is the initial phase. The symbol t is time duration, usually counted in seconds.

The angular velocity is sometimes expressed as $\omega = 2\pi f$, where f is the frequency, namely how many cycles of the phenomenon occur in one second. The period, which is the inverse of frequency, is $T = 1/f$.

The generation of sinewaves is most easily visualized in connection with motion around a circle. Figure 2-5 represents a point (P) moving around a circle (radius A) in a counterclockwise direction with angular velocity ω radians per second. The figure also shows the curves generated by the projection of the moving point on the horizontal x axis and the vertical y axis. These curves trace the cosine and sine respectively. In the construction, it is assumed that the initial

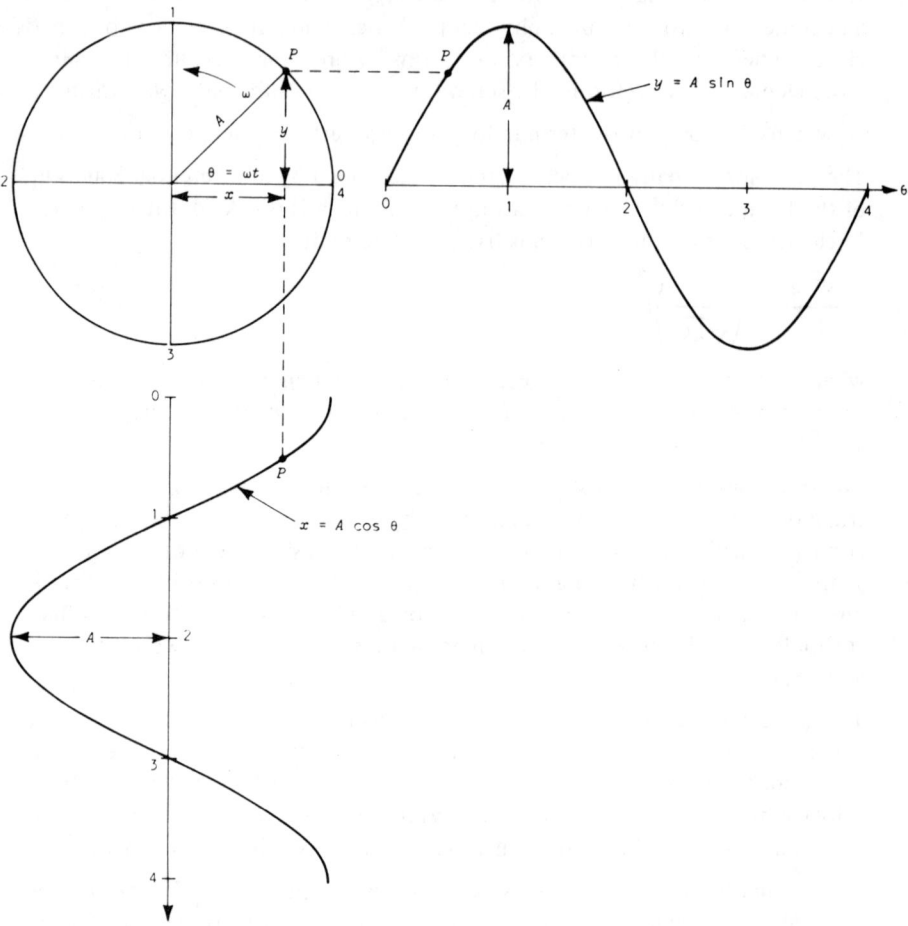

Figure 2-5 Generating Sinewaves by Circular Motion

position of P is that marked 0; if this were not the case, the starting angle α as an initial phase angle would be added.

Such a diagram is very useful since many of the characteristics of sinewaves can be deduced directly from it. For example, when the angle (θ) is zero, the horizontal projection is A while the vertical is zero. Likewise, when the angle is $\pi/2$ radians (90°), the horizontal projection is zero while the vertical is A. Going further, the two projected lengths of the radius A are equal when the point P is midway between zero and $\pi/2$. At this time the x and y projections

Spectrum Theory

are $A/\sqrt{2}$. A table of values for the sine and cosine is shown in Table 2-1. The table shows the angle θ in degrees; of course, radians could be used since one full circle is 2π radians or $360°$.

In Figure 2-5, at $270°$ the point (P) is at position 3. This position can also be reached by moving $90°$ clockwise. So, if counterclockwise rotation is positive and clockwise rotation is negative, $270°$ is equivalent to $-90°$ and vice versa. Using this type of notation and Table 2-1, the following relationships should hold: $\sin \theta = -\sin(-\theta)$, $\cos \theta = \cos(-\theta)$. Many other relationships can be developed in a similar manner.

The rotating point (P) starts at position zero at time $t = 0$ and starts moving counterclockwise. Eventually, (P) returns to its starting point (position 4); at this instant the elapsed time is some number, call it T. This time (T) is what was previously defined as the period. Likewise, the point (P) completes $f = 1/T$ complete trips around the circle every second. Since there are 2π radians in a circle, the angle that is covered in one second is $2\pi f$ (the angle per circle times circles per second). The angle swept to any arbitrary time t is merely the angle per second times the time in seconds or $\theta = 2\pi ft$. Thus our basic equations are $y = A \sin 2\pi ft$, $x = A \cos 2\pi ft$, where $2\pi f = \omega$, the angular velocity, and f is the frequency as previously defined.

Table 2-1

$\theta =$	0	$45°$	$90°$	$135°$	$180°$	$225°$	$270°$	$315°$	$360°$
$\sin \theta$	0	$1/\sqrt{2}$	1	$1/\sqrt{2}$	0	$-1/\sqrt{2}$	-1	$-1/\sqrt{2}$	0
$\cos \theta$	1	$1/\sqrt{2}$	0	$-1/\sqrt{2}$	-1	$-1/\sqrt{2}$	0	$1/\sqrt{2}$	1

Besides the familiar notation, such as $x = \cos(\omega t + \alpha)$, used above, it is possible to represent sinewaves in diverse ways. One way, which was used above, is by means of the rotating vector* which is the radius of the circle in Figure 2-5. The cosine function is represented by the projection of the rotating vector on the horizontal axis. The vector idea can be useful when dealing with the superposition† of several sinusoids. Thus, to find the sum of $A \cos \omega_0 t$ and $B \cos(\omega_0 t + \alpha)$, construct a vector triangle with two vectors size A and B forming the angle α. The resultant vector (or phasor), rotating at angular velocity ω_0, is the solution.

Strictly speaking, the single vector method is not quite correct since the rotating

---

*Such a vector is sometimes termed a phasor.
†Superposition is discussed more fully in Chapter 3. For the moment it is sufficient to equate superposition with addition.

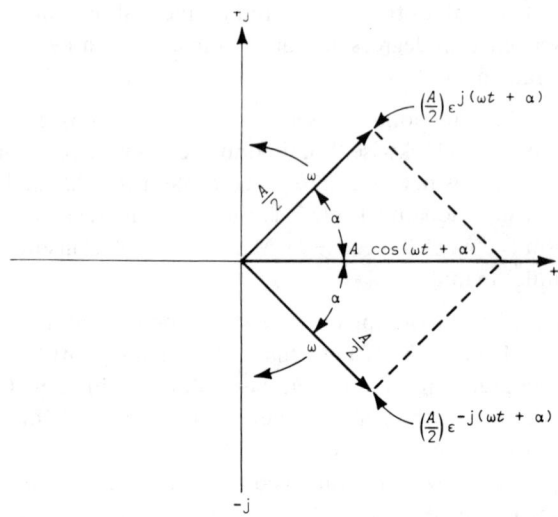

Figure 2-6 Representing the Cosine Function by Two Counter-Rotating Vectors on Complex Plane

vector generates both a sine and a cosine function. In order to represent only one of these, it is necessary to use two vectors rotating in opposite directions, as is illustrated in Figure 2-6. Here the two vectors have equal projections on the horizontal axis, which represents the cosine, but equal and opposite projections on the vertical axis, which represents the sine. Thus, if these vectors are half the amplitude of the single vector representation, by addition, the same cosine function is obtained as before, but the amplitude of the sine is zero since the two vectors cancel each other.

Such vectors can be represented in complex notation and manipulated in accordance with the rules of complex mathematics. Thus, in the complex plane, the horizontal axis is considered to represent real numbers while the vertical axis, labeled j, is considered to represent imaginary numbers. The word imaginary arises from $j = \sqrt{-1}$, which was once considered not to have any meaning, and is just a name representing a mathematical expression; it has nothing to do with existence or nonexistence. Complex notation provides a simple mathematical expression for a complex combination of sines and cosines; such notation is frequently used in Fourier series. In complex notation, a rotating vector is represented by the product of the amplitude and epsilon (the base of natural logarithms) raised to the power of the angle times j. In Figure 2-6, the two counter-rotating vectors are represented in complex notation by

Spectrum Theory

$$\left(\frac{A}{2}\right) \epsilon^{+j(\omega t + \alpha)} \quad \text{and} \quad \left(\frac{A}{2}\right) \epsilon^{-j(\omega t + \alpha)}$$

where the minus in front of the exponent on the second vector represents clockwise rotation, since counterclockwise is taken as the positive direction. The fact that such notation involves the use of negative frequencies should cause no concern. Remember, since it is a mathematical notation, it is not necessary to ascribe physical reality to all the parts. Another way of looking at it is that the positive and negative aspects are just different notation standing for the same thing, since, as shown before, $\sin(\theta) = -\sin(-\theta)$ and $\cos(\theta) = \cos(-\theta)$.

The relationships between the complex notation and standard trigonometric notation can be derived from the rules governing the use of complex numbers. The truth of the basic relationships can also be surmised with the help of a geometrical construction such as the one shown in Figure 2-6. Here it was shown that the sum of the two counter-rotating vectors is the cosine function. Thus,

$$\frac{A}{2}\left(\epsilon^{j(\omega t + \alpha)} + \epsilon^{-j(\omega t + \alpha)}\right) = A \cos(\omega t + \alpha)$$

From geometrical reasoning such as this come the following results:

$$\frac{\epsilon^{j\theta} + \epsilon^{-j\theta}}{2} = \cos\theta \tag{2-1}$$

$$\frac{\epsilon^{j\theta} - \epsilon^{-j\theta}}{j2} = \sin\theta \tag{2-2}$$

$$\epsilon^{j\theta} = \cos\theta + j\sin\theta \tag{2-3}$$

These are not independent expressions; Equation (2-1) can be derived from Equation (2-3):

$$\epsilon^{-j\theta} = \cos(-\theta) + j\sin(-\theta) = \cos\theta - j\sin\theta$$

$$\underline{\epsilon^{j\theta} \qquad\qquad\qquad = \cos\theta + j\sin\theta}$$

$$\epsilon^{-j\theta} + \epsilon^{j\theta} \qquad\qquad = 2\cos\theta$$

A sinusoid contains three basic parameters: amplitude, frequency, and initial phase. Since a graphical representation in the frequency domain is essentially a two-dimensional process, only two parameters can be represented per graph. Thus it takes two graphs to define all three of the parameters. This can be done in two ways. One is to present a graph of frequency and initial phase, while the

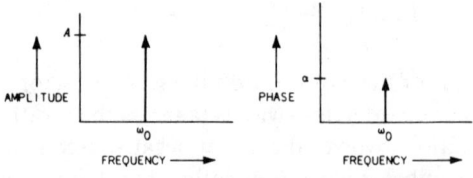

(A) Frequency-Phase Representation

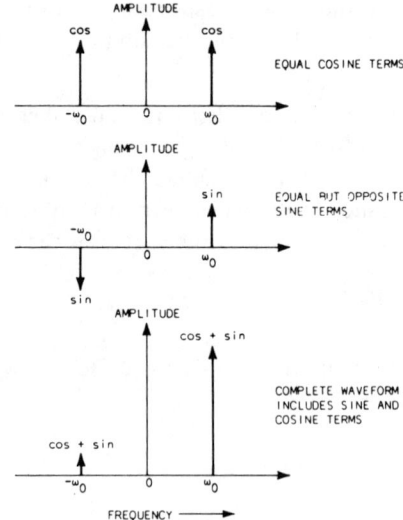

(B) Two Counter-Rotating Vectors Representation

Figure 2-7 Frequency Domain Representations of Sinusoid

other way is to use the two-vector representation showing a negative as well as positive frequency. The trigonometric expression can be reconstructed from either graph. Figure 2-7 shows both graphical representations. The complex notation, representing the two-vector technique, will be used for analytical purposes; the graphical methods will be strictly of the single positive-frequency type. This is because in the majority of spectrum analyzer problems, phase is of no interest. Thus all that is of interest can be represented in a single simple frequency-amplitude diagram. We have gone full circle back to Figure 2-1(B), except that instead of squarewaves, sinewaves as the basic function are utilized.

2.4. THE FOURIER SERIES

The sinusoidal constituents of a given waveform can be determined by various means. One particular method is the Fourier series. Other mathematicians had used this series prior to Fourier, but it was Fourier who, in a series of papers starting in 1822, showed the universal applicability of the series that now bears his name.

The Fourier series is based upon the fact that any function, which meets the three conditions stated below, can be expanded in a series of sines and cosines. The so-called Dirichlet conditions, that a function should meet in order to be Fourier expandable, are:

1. The function f(t) must have only a finite number of maxima and minima for the interval of definition.
2. The function must have only a finite number of finite discontinuities.
3. If the function has infinite discontinuities, its integral must be convergent — i.e.,

$$\int_{-\infty}^{+\infty} |f(x)| \, dx < N \text{ (a finite number)}$$

These conditions are easily met in practice, since no physical circuit can produce an infinite discontinuity. Even when such discontinuities are approached, such as in an approximate impulse function, the third condition is always met. Thus the Fourier series is highly suitable to the solution of practical problems.

The basic Fourier series is:

$$f(x) = \frac{1}{2} a_0 + (a_1 \cos x + b_1 \sin x) + (a_2 \cos 2x + b_2 \sin 2x) + (a_3 \cos 3x + b_3 \sin 3x) \ldots$$

It will be observed that this equation consists of a set of harmonically related terms where the lowest frequency terms (cos x, sin x) are considered the fundamental and all other terms are called the harmonics. A harmonic relationship means that all frequencies are integral multiples of the fundamental. Thus, if the fundamental is cos x, it is impossible to have a term of $\cos[(3/2)x]$. This makes computation easier; once the fundamental frequency is known, all others become obvious. There is, however, a drawback: More terms than might otherwise be necessary have to be used to describe adequately a particular function. For example, a function consists of cos t + 2 cos 1.5t. Since in Fourier series notation all terms have to be harmonics of some fundamental, the function would have to be expressed as 0 cos (.5t) + 1 cos 2(.5t) + 2 cos 3(.5t). Thus, even though the result is the same, since the first term is zero, it is nevertheless necessary to deal with three rather than two terms.

Though there are disadvantages, such as in the example above and the Gibbs phenomenon for discontinuous functions discussed in Chapter 3, the truncated Fourier series provides the closest approximation to an arbitrary function f(t).

In actual practice, an infinite number of terms cannot be dealt with, and it turns out that the Fourier series will give the least squared error approximation. (A demonstration of this appears at the end of this chapter.) Thus, even though in some instances this series has some disadvantages, it is the best general solution to practical problems.

2.5. DO SPECTRAL LINES EXIST?

According to Fourier theory, a squarewave consists of a fundamental sinewave, having the same period as the squarewave, and odd harmonics, whose amplitudes decrease in proportion to harmonic number. Mathematically, a squarewave is said to consist of

$$\frac{2Vt_0}{T} \left(\frac{1}{2} + \frac{2}{\pi} \cos \theta - \frac{2}{3\pi} \cos 3\theta \ldots \right)$$

as is discussed in more detail in Chapter 3. The frequency-domain representation along with that of a single sinusoid, $\cos \theta$, is shown in Figure 2-8. Few people have difficulty in visualizing the physical existence of the sinusoid represented in Figure 2-8(A). On the other hand, many have difficulty in visualizing the physical existence of the sinusoids, or spectral lines as these are often called, represented in Figure 2-8(B). "I know," the statement often goes, "that a squarewave can be treated as if it were made up of sinewaves, but do the sinewaves *really* exist?" This is a difficult philosophical question. Fortunately, it need not be resolved for the practical utilization of Fourier techniques because linear time-invariant circuits behave as if these sinewaves did in fact exist.

The fact that many time-variable circuits behave as if spectral lines did not exist is used as "proof" by those who choose not to believe in their existence. The believers, on the other hand, argue that all the above proves is that time-variable circuits are poor "detectors" of spectral lines. The fact that a person cannot detect light does not necessarily mean that there is no light. The same result occurs if the person has poor detectors, that is if the person is blind. What the question comes down to is the ancient one of primary and secondary qualities. The primary qualities are those that really exist in an object or phenomenon, while the secondary qualities are those that only seem to exist by virtue of interaction with the detector. This question has acquired modern importance in the area of quantum mechanics where the role of the observer is of the greatest significance.

The resolution of the question — Does thunder make a sound when there is no one there to hear it? — may have important philosophical implications, but it contributes very little to a practical discussion on spectrum analysis. The same may be said regarding the question on the *real* existence of spectral components. Rather, the question that should be asked is: Do real circuits behave as if spectral components exist?

Spectrum Theory 25

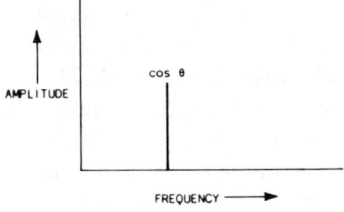

(A) Single Sinusoid

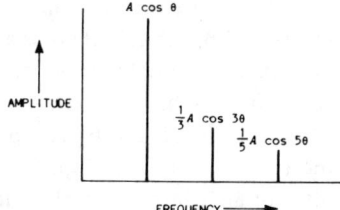

(B) Part of Series Forming Squarewave

Figure 2-8 Frequency Domain Representation (Phase Not Shown)

2.6. RESPONSE OF CIRCUITS TO SIGNALS

First it is necessary to emphasize that the term circuits means linear time-invariant circuits. Fourier theory does not necessarily apply to nonlinear or time-variable circuits. Though all circuit elements eventually become nonlinear, and no physical resistor will obey Ohm's law at infinitely large voltages, it is fortunate that for the range of signals of interest a majority of circuits are linear. What is of interest then is how linear time-invariant circuits will behave under the stimulus of an arbitrary signal input.

Generally, a circuit that is excited by an arbitrary input will respond in two distinct ways. One is called the *steady-state* response, while the other is the *transient* response. Most people, when confronted by the words steady state and transient think of a long time interval and a short time interval respectively. Though it is true that the transient state is usually characterized by a short time interval, this type of classification can be misleading, since there can be conditions where a steady state is never reached. Of greater importance to our discussion is the fact that transient behavior is determined by the so-called force-free solution, which means essentially that the basic characteristics of the transient response are determined completely by the circuit parameters. Thus, if a current pulse is injected into an R, L, C circuit formed into a loop, damped oscillations at radian frequency $\omega = 1/\sqrt{LC}$ will result when $(R/2L)^2 < 1/LC$. The shape and amplitude of the current pulse will help determine the amplitude of the oscillations, but the oscillating frequency is independent of the forcing function. This is a very important point because it leads to the following results. In order to test for the existence of spectral lines, a set of very narrowband contiguous filters is subjected to various inputs. What should be the result? Based on the previous discussion on the transient response, the output of each filter will be either nothing or a damped sinusoid at filter frequency; nothing else is possible. If the filter bandwidths are sufficiently narrow, the damping will be very slight and for all intents and purposes it will appear as if there is a continuous wave. Actually, the result could have been anticipated without any knowledge of transient behavior. Obviously, a filter can only have an output within its passband since that is what is meant by the word filter. But by computing the output amplitude distribution as a function of the type of input, there is the remarkable result that the transient response for an arbitrary input is identical to the steady-state response when the steady-state response is computed for an input composed of the Fourier components of the original arbitrary input.* Hence — real, linear, time-invariant circuits behave as if spectral lines did in fact exist.

If one chooses to believe in the real existence of spectral lines, then a Fourier analysis is simply a computation of some of the parameters of a signal. These parameters are eventually used in the practical business of determining the steady-state response of some network. If, on the other hand, one chooses not to believe in the real existence of spectral lines, then a Fourier analysis is simply a mathematical procedure that has no counterpart in physical reality. It is just a convenient technique for solving problems — similar, for example, to the laying out of an orderly array of numbers in determinants when solving simultaneous equations. The solution of the equations may correspond to something physical, but it certainly is not necessary to validate the technique of solution by finding some entity that is physically spread out in an orderly array similar to the determinant.

The great utility of the Fourier technique is that it permits the solution of complicated transient problems by relatively simple steady-state techniques. This alone is sufficient justification for its use. If, in addition, one believes in the existence of spectral lines, then the advantages of Fourier techniques are obvious.

All the arguments advanced for the use of Fourier techniques also hold true for the use of the instrument called a spectrum analyzer. Again one can look at this

---

*See Weber, *Linear Transient Analysis,* Vol. II, for a discussion on the response of ideal filters to pulses.

Spectrum Theory

in two ways. One view is that the display shows the spectral or energy distribution of the signal. The second view is that the display is the transient response of the spectrum analyzer circuits in response to the stimulus of the signal. Both views lead to the same final result: The spectrum analyzer display can be used to compute or predict the response of various linear time-invariant circuits under the same stimulus. This is the only justification necessary for the use of this instrument.

2.7. PROOFS AND EXAMPLES

2.7.1. Orthogonal Functions

The series 1, x, $x^2 - (1/3)$, $x^3 - (3/5)x$ ... is a series of orthogonal functions between the limits ± 1.

For any series of orthogonal functions:

$$\int_a^b f_m(x) f_n(x) \, dx = 0 \qquad m \neq n$$

For the above series:

$$\int_{-1}^{+1} (1)(x) \, dx = \left[\frac{1}{2} x^2 \right]_{-1}^{+1} = \frac{1}{2} - \frac{1}{2} = 0$$

$$\int_{-1}^{+1} (1)\left(x^2 - \frac{1}{3}\right) dx = \left[\frac{1}{3} x^3 - \frac{1}{3} x \right]_{-1}^{+1} = 0$$

$$\int_{-1}^{+1} (x)\left(x^2 - \frac{1}{3}\right) dx = \left[\frac{1}{4} x^4 - \frac{1}{6} x^2 \right]_{-1}^{+1} = 0$$

$$\int_{-1}^{+1} (1)(1) \, dx = \left[x \right]_{-1}^{+1} = 2$$

$$\int_{-1}^{+1} (x)(x) \, dx = \left[\frac{1}{3} x^3 \right]_{-1}^{+1} = \frac{2}{3}$$

$$\int_{-1}^{+1} \left(x^2 - \frac{1}{3}\right)\left(x^2 - \frac{1}{3}\right) dx = \left[\frac{1}{5} x^5 - \frac{2}{9} x^3 + \frac{1}{9} x \right]_{-1}^{+1} = \frac{8}{45}$$

which shows that the integral of the product of different terms is zero while the integral of a term times itself is not zero. Hence, the series is composed of a set of orthogonal functions.

2.7.2. Complex Notation

The Euler identity

$$e^{j\theta} = \cos\theta + j\sin\theta$$

can be used to compute the values of various complex expressions. Thus,

$$e^{j\pi} = \cos\pi + j\sin\pi = -1 + j0 = -1$$

Similarly, the reader can verify that

$$e^{j(\pi/2)} = +j$$

$$e^{-j(\pi/2)} = -j$$

$$e^{-j(3\pi/2)} = +j$$

$$e^{j2\pi} = +1$$

Besides the trigonometric form, complex quantities can also be expressed in algebraic form. Thus $Z = a + jb$ is a complex quantity, $Z^* = a - jb$ is called the conjugate of Z.

A complex quantity and its conjugate have a specific relationship to each other. For example, their sum is real;

$$Z + Z^* = (a + jb) + (a - jb) = 2a$$

Likewise, keeping in mind that $j = \sqrt{-1}$,

$$Z \cdot Z^* = a^2 + b^2$$

$$Z - Z^* = j2b$$

2.7.3. Proof that the Truncated Fourier Series Provides a Least Squared Error Fit

Let f(t) be an arbitrary function of time that is to be expanded over some specified range in an infinite series of sines and cosines. If we take only a finite number of terms, then the series is only approximately equal to f(t). Thus:

$$f(t) \cong \sum_{m=0}^{n} [a_m \cos m\omega t + b_m \sin m\omega t]$$

where the terms being summed are an arbitrary collection of sines and cosines and not necessarily a Fourier series. To show that in order to have the least squared error between f(t) and the nonfinite series, which stops at m = n, approximating f(t), it is necessary that the series be a Fourier series. Thus, the squared error is

$$\rho = \frac{1}{T} \int_{-T/2}^{+T/2} [f(t) - \Sigma]^2 dt$$

where the sigma (Σ) is a shorthand notation for the truncated series. In order for the error (ρ) to be minimum, the partial derivative of the error with respect to any coefficient should be zero. Thus, to find the minimum error with respect to the coefficients of the cosine terms (a_m), the derivatives should be equated to zero:

$$\frac{\partial \rho}{\partial a_m} = \frac{1}{T} \int_{-T/2}^{T/2} \frac{\partial}{\partial a_m} [f(t) - \Sigma]^2 dt$$

$$= \frac{1}{T} \int_{-T/2}^{T/2} 2[f(t) - \Sigma](-\cos m \omega t) dt = 0$$

Separating the terms:

$$\int_{-T/2}^{T/2} f(t) \cos m \omega t \, dt = \int_{-T/2}^{T/2} (\Sigma) \cos m \omega t \, dt$$

The summation sign, of course, stands for

$$\sum_{m=0}^{n} [a_m \cos m \omega t + b_m \sin m \omega t]$$

Now the definite integral, taken over one period, of a sine-cosine product is zero. Likewise, by virtue of orthogonality, are all cosine-cosine products except those where the two terms are the same. Hence

$$\int_{-T/2}^{T/2} f(t) \cos m \omega t \, dt = \int_{-T/2}^{T/2} a_m \cos^2 m \omega t \, dt = \frac{a_m T}{2}$$

Solving for a_m:

$$a_m = \frac{1}{T/2} \int_{-T/2}^{T/2} f(t) \cos m \omega t \, dt$$

Likewise, it can be shown that

$$b_m = \frac{1}{T/2} \int_{-T/2}^{T/2} f(t) \sin m \omega t \, dt$$

which are the same as the Fourier coefficients, hence the Fourier series gives the best fit.*

2.7.4. Example of Time-Variable Network

Consider the system illustrated in Figure 2-9. Suppose the input is a single continuous sinewave at a frequency within the passband of the filter and the switch

---

*A proof showing identity between individual terms will be found in Weber, *Linear Transient Analysis*, Vol. I, pages 258-259.

is closed. Obviously, the indicator shows a response. Now leaving the switch open for a short period of time during each cycle, the indicator may read something different than before, but the indicator continues to show an output.

Let the input be changed to a train of narrow pulses having the same period as the sinewave. According to Fourier theory, the pulse train can be considered as the sum of an infinite number of sinewaves. One of these Fourier sinewaves, which has the special name of fundamental, should be at the same frequency as the original sinewave since the pulse train has the same period as the original sinewave.

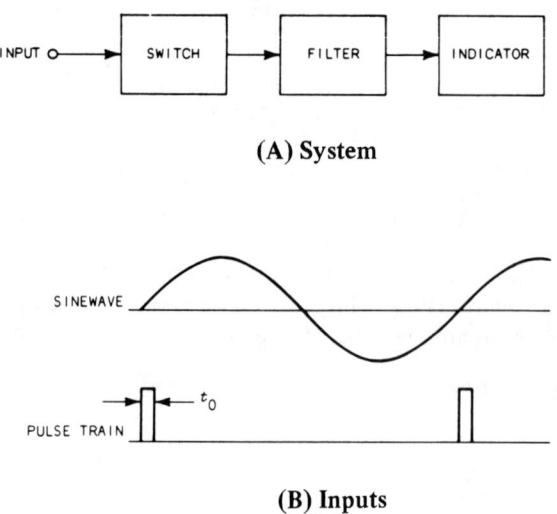

(A) System

(B) Inputs

Figure 2-9 Time Variable Network

The switch is closed. There is a response on the indicator, which means the fundamental is there. Now leave the switch open for a short period of time (t_0) during each cycle, so arranged that the switch opening coincides with the occurrence of a pulse. Unlike the case of the single sinewave, the indicator now shows nothing. Obviously, the results differ for both inputs. The reason for this is that the switch makes this a time-variable network for which Fourier theory does not apply.

Spectrum Theory

2.8. EXERCISES

2-1. Which of the pairs of functions are orthogonal between the limits of ±2?
- a. $1, X$
- b. $2X, X^2$
- c. $3X, X^3$
- d. $4X, X^4$

2-2. Verify that:

a. $\dfrac{e^{j\pi} + e^{-j\pi}}{2} = e^{j\pi} = -1$

b. $e^{j(\pi/2)} = +j$

c. $e^{-j(\pi/2)} = -j$

d. $e^{j2\pi} = +1$

2-3. Which of the following cannot be a Fourier series?

a. $X - \dfrac{X^2}{2} - \dfrac{X^3}{3} + \dfrac{X^4}{4}$

b. $5 \sin X + 7 \sin 3X - 2 \sin 4X + 1 \sin 5X$

c. $\sin X - 2 \cos 3X + 4 \sin 4X + \cos 4X$

d. $\sin X + 3 \sin 2.7X - \cos \pi X + 2 \tan \pi X$

Chapter 3
Fourier Analysis

3.1. INTRODUCTION

All practical functions defined over an interval, such as $-\pi$ to $+\pi$, can be expanded in a Fourier series. This holds true even when the waveform represented by the function is nonrepetitive and exists only during the defined interval. The interpretation of the Fourier series of these isolated pulses has no meaning outside the defined interval. Single pulses can also be treated from the continuous spectrum Fourier integral point of view, and this is the approach that will be taken here. Thus in this text Fourier series will be considered for periodic functions only.

3.2. FOURIER SERIES

Practical, physically realizable functions having a period of 2π can be expanded in a series of trigonometric functions such that the function

$$f(x) = \frac{a_0}{2} + (a_1 \cos x + b_1 \sin x) + (a_2 \cos 2x + b_2 \sin 2x) \ldots$$
$$+ (a_n \cos nx + b_n \sin nx) \tag{3-1}$$

which can be represented in summation form

$$f(x) = \frac{a_0}{2} + \sum_{n=1}^{\infty} (a_n \cos nx + b_n \sin nx)$$

where x is an angle in radians.

The series defined by Equation (3-1) is known as a Fourier series. The coefficients a_n and b_n are constants that are determined by the form of the original function f(x). The two summations, sine and cosine, can be combined into a single series by the addition of a phase angle. Thus,

$$f(x) = \frac{C_0}{2} + \sum_{n=1}^{\infty} C_n \cos(nx + \phi_n) \tag{3-2}$$

where

$$C_0 = a_0$$

$$C_n = \sqrt{a_n^2 + b_n^2}$$

$$\phi_n = \tan^{-1}\left(\frac{-b_n}{a_n}\right) \qquad (3\text{-}2)$$

Equation (3-2) is the more useful for spectrum analyzer work, since the spectrum analyzer displays the combined amplitude, C_n, rather than the separate sine and cosine amplitudes. Furthermore, spectrum analyzers of the type under discussion do not display any phase characteristics. Therefore, except in special cases such as the combination of several complicated spectra, the phase angle, ϕ, will be ignored.

As is demonstrated at the end of this chapter, the coefficients a_n and b_n are related to the original function f(x) through the following integrals.

$$a_0 = \frac{1}{\pi}\int_{-\pi}^{+\pi} f(x)\, dx = \frac{1}{\pi}\int_0^{2\pi} f(x)\, dx$$

$$a_n = \frac{1}{\pi}\int_{-\pi}^{+\pi} f(x) \cos nx\, dx = \frac{1}{\pi}\int_0^{2\pi} f(x) \cos nx\, dx \qquad (3\text{-}3)$$

$$b_n = \frac{1}{\pi}\int_{-\pi}^{+\pi} f(x) \sin nx\, dx = \frac{1}{\pi}\int_0^{2\pi} f(x) \sin nx\, dx$$

As indicated in Equation (3-3) above, the integration can be carried out between $+\pi$ and $-\pi$ or between 0 and 2π. The choice depends on how the original function f(x) is defined and which integration involves less work. In any event, the final result is the same, regardless of which integration is used. Very often the period is some arbitrary time interval T rather than 2π. The Fourier series still applies except that all expressions have to be scaled by the factor $2\pi/T$; thus,

$$f(x) = \frac{a_0}{2} + \sum_{n=1}^{\infty}\left(a_n \cos 2n\frac{\pi}{T}x + b_n \sin 2n\frac{\pi}{T}x\right) \qquad (3\text{-}4)$$

and the coefficients are given by

$$a_n = \frac{2}{T}\int_{-T/2}^{+T/2} f(x) \cos \frac{2n\pi x}{T}\, dx = \frac{2}{T}\int_0^T f(x) \cos \frac{2n\pi x}{T}\, dx \qquad (3\text{-}5)$$

$$b_n = \frac{2}{T}\int_{-T/2}^{+T/2} f(x) \sin \frac{2n\pi x}{T}\, dx = \frac{2}{T}\int_0^T f(x) \sin \frac{2n\pi x}{T}\, dx$$

Fourier Analysis

Again, a_n and b_n can be combined into a single amplitude factor $C_n = \sqrt{a_n^2 + b_n^2}$, which is the factor displayed on the spectrum analyzer.

When using the C_n representation, it is still necessary to solve two separate equations, one for the a_n terms and the other for the b_n terms. These two equations can be combined into one by the use of complex notation.

Thus, in complex notation,

$$f(x) = \sum_{n=-\infty}^{n=+\infty} d_n e^{jnx} \tag{3-6}$$

where the coefficient d_n is obtained from

$$d_n = \frac{1}{2\pi} \int_{-\pi}^{+\pi} f(x) e^{-jnx} dx \tag{3-7}$$

and n takes on all positive and negative values as well as zero. Thus, except for the case where n = 0, there are still two terms for every n, namely d_n and d_{-n} with one being the conjugate of the other. The two complex coefficients, when combined in accordance with the rules for complex numbers (as discussed in Chapter 2), leads back to the more familiar trigonometric expression

$$d_n e^{jnx} + d_{-n} e^{-jnx} = \frac{a_n - jb_n}{2} e^{jnx} + \frac{a_n + jb_n}{2} e^{-jnx} \tag{3-8}$$

$$= a_n \cos nx + b_n \sin nx$$

The complex notation presents some conceptual difficulty because of the appearance of what seems to be negative frequencies, since the summation goes over negative as well as positive numbers. Some ideas on the interpretation of negative n's will be found in Chapter 2. On the other hand, the complex Equations (3-6) and (3-7) are much more compact than their trigonometric counterpart. The complex notation is particularly useful in the Fourier integral representation, which is needed for the analysis of continuous spectra.

3.3. FOURIER APPLICATIONS

In a Fourier series representation, the fundamental waveform is the sinusoid. Hence, it is of interest to determine the frequency-domain representation, by way of Fourier series, for a sinusoid. The time-domain function is $f(t) = A \cos \omega_0 t$. The frequency-domain function is the Fourier series previously defined. Using the trigonometric representation, Equation (3-5) is used to determine the Fourier coefficients a_n and b_n. Equation (3-5) is reproduced here as Equation (3-10):

$$a_0 = \frac{2}{T} \int_0^{+T} f(x) \, dx \tag{3-9}$$

$$a_n = \frac{2}{T} \int_0^{+T} f(x) \cos \frac{2n\pi x}{T} dx \qquad (3\text{-}10)$$

$$b_n = \frac{2}{T} \int_0^{+T} f(x) \sin \frac{2n\pi x}{T} dx$$

Just as the solution for the coefficients ($a_0 = 0$, $a_1 = \pi/\pi = 1$) is simple, demonstrating that the cosine can be represented by a delta function is difficult. This difficulty is because of the continuous cosine wave, which in engineering language is called a CW signal and is assumed to exist forever. This leads to the integration of sinusoids with infinite limits, which are normally not defined but can be handled via the theory of distributions.*

This solution can also be obtained by complex representation such as was discussed in Chapter 2, where it was shown that the cosine function can be represented by two in-phase impulse functions. The analytical expression for the graph of Figure 2-7 is given in Equation (3-11).

$$f(t) = A \cos \omega_0 t$$

$$\omega_0 = 2\pi f_0 t$$

$$F(\omega) = \frac{A}{2} \left[\delta(f + f_0) + \delta(f - f_0) \right] \qquad (3\text{-}11)$$

The function $\delta(x)$ is called a Dirac, delta, or impulse function, where $\delta(f)$ is a frequency impulse and $\delta(t)$ is a time impulse. The reason for the two parts in the Fourier representation is that in the solution of problems by the complex notation method, two coefficients are obtained; these are d_n and d_{-n} or in this case d_{+f0} and d_{-f0}. Since in practical applications one is interested in the composite amplitude, one might as well use the notation $A \delta(f - f_0)$, meaning an impulse at frequency $f = f_0$ and of strength A. An impulse has a finite area, in this case equal to A, and zero width, thus requiring infinite amplitude. Obviously, the impulse is only a theoretical function since infinite amplitudes cannot be generated. This does not mean that the impulse does not have validity. The impulse function is just as valid for analysis and theoretical representation purposes as the sinusoid of infinite time duration whose frequency representation it is. Certainly infinite-duration sinusoids do not exist in practice, but that doesn't prevent using these to represent practical signals. Some properties of impulse functions will be found in the appendix for this chapter.

The graphical representation of the impulse is symbolic rather than exact. This is because there is no way to faithfully graph a function that calls for infinite amplitude. A symbolic rather than exact graphical representation may disturb some people, but actually this is often done. The infinite-duration sinewave, for

---

*See Papoulis, *The Fourier Integral and its Applications*.

Fourier Analysis

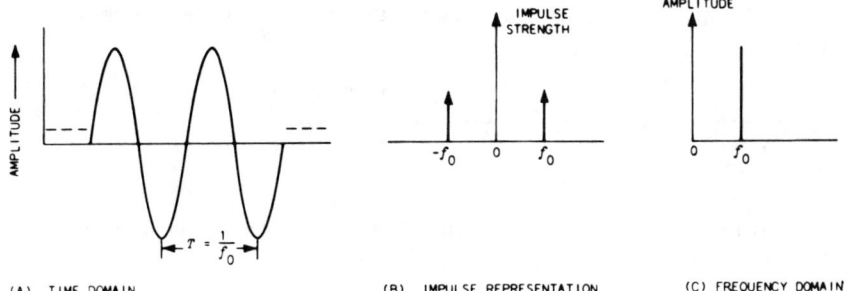

(A) TIME DOMAIN (B) IMPULSE REPRESENTATION (C) FREQUENCY DOMAIN

Figure 3-1 Time and Frequency Domain Representations For Sinusoid

example, is represented symbolically by a finite number of cycles rather than the infinite number of cycles that it is supposed to have. In any event, what is of interest in the case of the impulse is not its amplitude, which is always considered infinite, but rather the area or strength, which is representative of the amplitude of its time-domain sinewave equivalent. Figure 3-1 shows the time and frequency-domain representations of a simple CW signal, namely the sinewave and impulse. Figure 3-1(C) represents a spectrum analyzer display. Note the identity between the Fourier series derived in Figure 3-1(C) and the desired representation shown in Figure 2-8(A).

Let us consider now the important case of a rectangular pulse train of arbitrary pulse width (t_0) and arbitrary period (T) as shown in Figure 3-2(A).

The function f(x) is defined by the amplitude A for $-(t_0/2) < t < +(t_0/2)$ and zero everywhere else within the interval of integration. Using the scale factor $2\pi/T$ in Equation (3-7), we have

$$d_n = \frac{1}{T} \int_{-T/2}^{+T/2} f(t) \, \epsilon^{-jn(2\pi t/T)} \, dt \tag{3-12}$$

which is the complex notation equivalent of Equation (3-5).

Since the pulse only exists between the limits of $\pm (t_0/2)$, we only need to integrate between these limits. Thus,

$$d_n = \frac{1}{T} \int_{-(t_0/2)}^{+(t_0/2)} A \, \epsilon^{-jn(2\pi t/T)} \, dt = \frac{A}{T} \left[\frac{1}{-jn\frac{2\pi}{T}} \epsilon^{-jn(2\pi t/T)} \right]_{-t_0/2}^{+t_0/2}$$

By substituting the limits:

$$d_n = \frac{A}{T} \left(\frac{\epsilon^{-jn(2\pi/T)(t_0/2)} - \epsilon^{+jn(2\pi/T)(t_0/2)}}{-jn\frac{2\pi}{T}} \right)$$

By rearranging terms and bringing the minus sign from the denominator to the numerator:

$$d_n = \frac{A}{\frac{n\pi T}{T}} \left(\frac{\epsilon^{+jn(2\pi/T)(t_0/2)} - \epsilon^{-jn(2\pi/T)(t_0/2)}}{2j} \right)$$

The part in the brackets is the sine function as discussed in the appendix to Chapter 2. Hence,

$$d_n = \frac{A}{n\pi} \sin \frac{n\pi t_0}{T}$$

The same result is obtained for the d_{-n} components.

What is of interest is the overall amplitude term C_n, which is obtained by summing d_n and d_{-n}. Thus, after adding and rearranging terms, the final result is:

$$C_n = \frac{2At_0}{T} \frac{\sin \frac{n\pi t_0}{T}}{\frac{n\pi t_0}{T}} \qquad (3\text{-}13)$$

This is the origin of the so-called sine x over x (sin x/x) distribution, which will be used extensively in connection with continuous spectra.

The interpretation of Equation (3-13) is that the pulse train is made up of a dc component C_0 and a set of sinusoids with amplitudes $C_1, C_2, \ldots C_n$. The term C_0 is usually ignored in graphical representations; it does not show up on the spectrum analyzer. Therefore, a graph of a frequency distribution would consist of a representation of a train of impulses, where each impulse represents a sinusoid of the series. A sinusoid, as indicated in the previous example, is represented in the frequency domain by an impulse. However, it is the area rather than the amplitude of the impulse that is equivalent to the amplitude of the sinusoid.

A graphical display of a Fourier series with coefficients, such as those of Equation (3-13), therefore consists of a set of vertical lines that are symbolic impulses. One impulse is used per sinusoid in the series. As an example, let $T = 2t_0$, making the pulse train into a squarewave as shown in Figure 3-2(B). Substituting $T = 2t_0$ into Equation (3-13):

$$C_n = A \frac{\sin \frac{n\pi}{2}}{\frac{n\pi}{2}} \qquad (3\text{-}14)$$

Fourier Analysis

as the equation for the amplitudes of the sinusoids, which when combined make up the squarewave. The individual amplitudes are obtained by substituting for n. Thus, for the fundamental, n = 1:

$$C_1 = A \frac{\sin \frac{\pi}{2}}{\frac{\pi}{2}} = \frac{2A}{\pi} \quad (\text{since } \sin \pi/2 = 1)$$

For the second harmonic, amplitude n = 2 is substituted, leading to:

$$C_2 = A \frac{\sin \pi}{\pi} = 0 \quad (\text{since } \sin \pi = 0)$$

When a similar procedure is followed for the other harmonics, it is observed that all the even harmonics are zero, while the amplitude of the odd harmonics progresses as $1/n$, so that the fifth harmonic is one-fifth as large as the fundamental while the ninth harmonic amplitude is one-ninth as large as the fundamental, etc. In addition, there is also the dc or average term at n = 0. This term is difficult to obtain directly from Equation (3-14), since substitution of n = 0 leads to the indeterminant zero over zero. However, this term is easily obtained by looking at Figure 3-2(B); it is simply t_0/T or 1/2 for a squarewave.

The complete Fourier series representation for a squarewave, therefore, is

$$f(t) = A \left(\frac{1}{2} + \frac{2}{\pi} \cos \omega_0 t - \frac{2}{3\pi} \cos 3\omega_0 t \ldots \right) \quad (3\text{-}15)$$

where $\omega_0 = 2\pi/T$.

It should be kept in mind that although Equation (3-15) shows a dc term, the spectrum analyzer will not show this. Although Equation (3-14) shows alternating 180° phase reversals as indicated by the alternating plus and minus signs, the spectrum analyzer will not show this either. Figure 3-2(C) shows the frequency-domain characteristics of the squarewave.

3.4. SUPERPOSITION

One of the most important concepts that is applicable to Fourier analysis is that of superposition. The superposition principle states that the response of a linear network to an arbitrary periodic input function is equal to the sum of the responses of the constituents of the input function. When the individual components are expressible mathematically by ordinary algebraic functions, simple addition is all that is needed. When the signal components are vectors represented by complex notation, then the rules for vectorial addition, which take into account phase relationships, must be used.

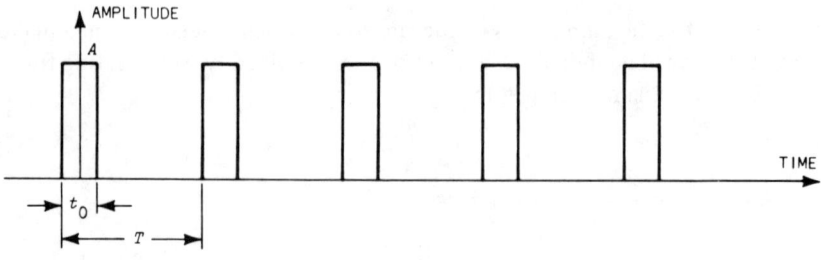

(A) Rectangular Pulse Train

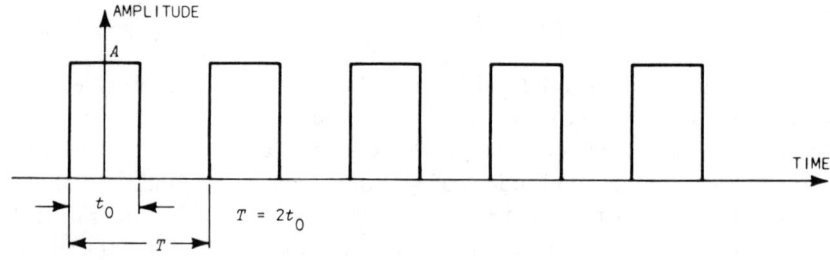

(B) Squarewave

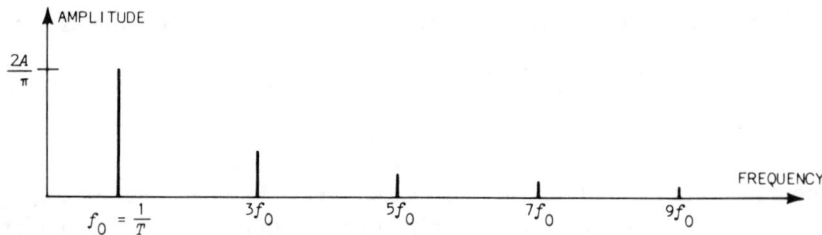

(C) Frequency Domain of Squarewave; dc Term and Phase Ignored

Figure 3-2 Time and Frequency Domain Representations of a Pulse Train

Superposition is tacitly assumed in the formulation of Fourier series. This is because the statement that a complex waveform can be represented by a sum of sinewaves cannot be made unless superposition holds so that the sinewaves can be summed. A more significant use of superposition is in the relationship of cause and effect. Thus, in using Fourier series in the solution of network response problems, the solution is frequently obtained by performing a Fourier analysis of the input signal and then summing the network responses obtained by considering each of the Fourier sinewave components as being applied to the network singly. Of more importance in spectrum analysis is the fact that the spectra of complex waveforms can be obtained by superposition. Thus, if the

Fourier Analysis

Fourier representation for a time-domain function f(t) is F(f) and for another time-domain function g(t) it is G(f), then the Fourier representation for the combined time-domain function f(t) + g(t) is just F(f) + G(f). This, when combined with the fact that a time delay introduces a phase shift but otherwise leaves the spectrum unaffected, permits the computation of complex spectra simply by the addition of simpler spectra. Thus, for example, the frequency-domain characteristics of a trapezoid can be found by the addition of the spectra of two triangular pulses and one rectangular pulse.

As a specific example, consider a squarewave as constructed by the addition of two sawtooth waves as shown in Figure 3-3. This squarewave is identical to that shown in Figure 3-2(B) except for the elimination of the dc term. The Fourier coefficients for a sawtooth, which can be obtained by the standard method of integration, are given by $C_n = A(1/\pi n)$, where A is the maximum height (shown as 2 in Figure 3-3).

The sum of the two Fourier series is

$$f(\theta) = \frac{2}{\pi} \left[\frac{1}{2} - \sin(\theta - \pi) - \frac{1}{2} \sin 2(\theta - \pi) - \frac{1}{3} \sin 3(\theta - \pi) \ldots \right]$$
$$- \frac{2}{\pi} \left[\frac{1}{2} - \sin \theta - \frac{1}{2} \sin 2\theta - \frac{1}{3} \sin 3\theta \ldots \right] \qquad (3\text{-}16)$$

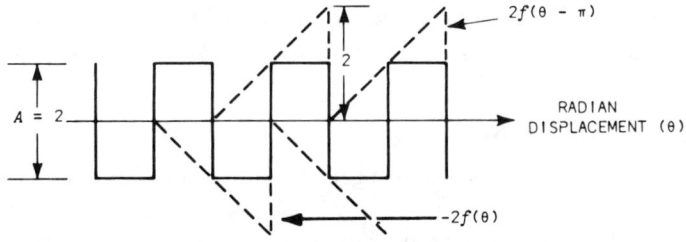

Figure 3-3 Squarewave as Sum of Two Sawtooth Waves

The dc terms of the sawtooth waves are equal and opposite and hence cancel. The bottom sawtooth has a half-cycle phase shift with respect to the top one, hence the $(\theta - \pi)$ term.

The two sinewave series in Equation (3-16) can be combined into a single series with the help of the trigonometric identity:

$$\sin n (x - \pi) = (-1)^n \sin (nx) \qquad (3\text{-}17)$$

Using Equation (3-17), Equation (3-16) becomes:

$$f(\theta) = \frac{2}{\pi}\left(\frac{\sin\theta}{1} + \frac{\sin 2\theta}{2} + \frac{\sin 3\theta}{3} \cdots\right)$$
$$-\frac{2}{\pi}\left(-\frac{\sin\theta}{1} + \frac{\sin 2\theta}{2} - \frac{\sin 3\theta}{3} \cdots\right) \tag{3-18}$$

The series in Equation (3-18) can be added term by term resulting in

$$f(\theta) = \frac{4}{\pi}\left(\frac{\sin\theta}{1} + \frac{\sin 3\theta}{3} + \frac{\sin 5\theta}{5}\right) \tag{3-19}$$

Except for the loss of the dc term because of the vertical shift of the squarewave, the use of θ instead of ωt to simplify the notation, and the phase shift resulting from considering zero time at the start rather than in the middle of a cycle as in Figure 3-2, Equations (3-15) and (3-19) are the same. Actually, the analysis could have easily been arranged so that Equations (3-15) and (3-19) would be identical.

Besides illustrating the use of the principle of superposition, it is the intent to illustrate that, regardless of how the analysis is performed, the essential features which are those that are displayed on a spectrum analyzer, remain the same. The spectrum analyzer does not display the dc term, nor is the spectrum analyzer sensitive to phase, so that the switch from cosine to sine has no effect. The important part of the analysis is that a squarewave is represented by an infinite series of sinusoids consisting of a fundamental and odd harmonics, with harmonic amplitude decreasing as the inverse of the harmonic number. This is precisely the information that can be obtained by means of the spectrum analyzer.

3.5. GIBBS PHENOMENON

Based on the previous discussion, one gets the impression that one should be able to reconstruct any waveform simply by adding the sinusoids forming the Fourier series. Of course, since most Fourier series call for an infinite number of terms, such a reconstruction is not practical.

Nevertheless, this can be considered theoretically. Such a study helps in establishing the validity of the Fourier approach and is useful in establishing guidelines on how many terms of the series can be considered a sufficiently close approximation. When this is done for a discontinuous function, such as a squarewave, the result is not that shown in Figure 3-2(B) but rather that shown in Figure 3-4.

As the number of harmonics in the summation is increased the resultant waveform is seen to oscillate around the discontinuities at the corners as shown in Figure 3-4(A). This occurs because in the vicinity of a finite discontinuity, the sum of the Fourier terms converge to the average value as the discontinuity is

Fourier Analysis

approached from both sides. As the number of terms in the series is increased, the oscillation squeeze closer and closer together; and as the number of terms approaches infinity, the oscillation is squeezed into a straight line as shown in Figure 3-4(B). This oscillating overshoot phenomenon is known as the Gibbs phenomenon in Fourier series. There is no way to get away from this overshoot when summing a Fourier series except by modifying the coefficients, in which case, it is of course no longer a straightforward Fourier series.*

The size of the overshoot, as the limit in the number of terms in the series is increased without bound, has been computed by many people including Weber, *Linear Transient Analysis,* vol. I. Under the best circumstances, the overshoot is about 18%, as shown in Figure 3-4(B).

The fact that the sum of the Fourier terms does not seem to lead back to the original function appears, at first glance, to be a serious blow to Fourier theory. Actually, from an energy point of view, there is no discrepancy. This is because the overshoot is an infinitely thin line of finite amplitude and, hence, has zero area and no energy. It can of course be argued that infinitely thin lines do not

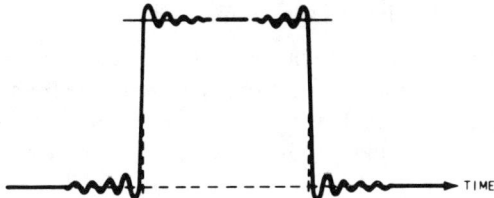

(A) Reconstructed Squarewave, Finite Number of Harmonics

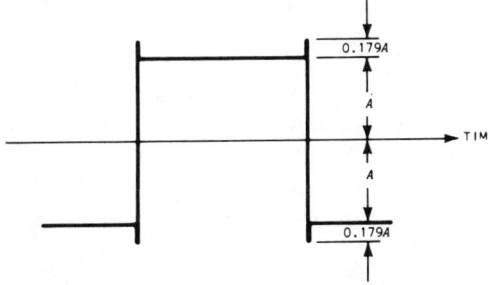

(B) Reconstructed Squarewave as Number of Harmonics Goes to Infinity

Figure 3-4 Squarewave Reconstructed from Fourier Components

---

*A discussion on this and other aspects of the Gibbs phenomenon will be found in Guillemin's *The Mathematics of Circuit Analysis.*

exist in practice, but then neither do the infinitely steep slopes of perfect squarewaves. What it eventually comes down to is the question of how real physically realizable circuits behave, which is discussed in Chapter 2. It might be well to repeat the major point on which all spectrum analyzer work is based: Physically realizable, linear, time-invariant networks behave as if Fourier spectral components exist. The function of the spectrum analyzer is to provide information on the behavior of such circuits. Therefore, it is not necessary to believe in the "real" existence of Fourier spectral lines in order to accept the validity of the spectrum analyzer display.

3.6. CONTINUOUS — DENSE SPECTRUM

While the concept of a frequency spectrum for a pulse train is at least intuitively acceptable, the concept of a frequency spectrum for a single pulse is much more difficult to comprehend. Indeed the frequency distributions cannot be treated in the same manner; the former is a discrete or line-type spectrum while the latter is a continuous dense type of spectrum. The mathematical treatment for these two types of spectra is also different, since the Fourier series is not directly applicable to the continuous spectrum.

The continuous spectrum is handled easiest when considered as the limiting case of a discrete spectrum. Let us, therefore, start with the frequency-domain representation of a pulse train such as that shown in Figure 3-2(A). The Fourier coefficients are given by Equation (3-13), which is reproduced in Equation (3-20).

$$C_n = \frac{2At_0}{T} \frac{\sin \frac{n\pi t_0}{T}}{\frac{n\pi t_0}{T}} \tag{3-20}$$

Disregarding the dc term, which will not be displayed on the spectrum analyzer anyway, one obtains the amplitude of the fundamental and the various harmonics by substituting $n = 1, 2, 3 \ldots$ into Equation (3-20). Thus, the amplitude of the fundamental is

$$C_1 = \frac{2At_0}{T} \frac{\sin \frac{\pi t_0}{T}}{\frac{\pi t_0}{T}} \tag{3-21}$$

The second harmonic amplitude is

$$C_2 = \frac{2At_0}{T} \frac{\sin \frac{2\pi t_0}{T}}{\frac{2\pi t_0}{T}}$$

and so forth. The important and interesting case occurs when t_0/T is small; in other words, when we are dealing with a train of narrow pulses rather than squarewaves. Under such conditions the angle $(\pi t_0/T)$ is small and Equation (3-21) reduces to $C_1 = 2At_0/T$ because the sine of a small angle is essentially equal to the angle. This approximation only holds true in the vicinity of the fundamental; obviously the angle $(n\pi t_0/T)$ eventually gets large as the harmonic number (n) increases. As the harmonic number is increased, the expression represented by Equation (3-20) decreases. Eventually, a point is reached where $n = T/t_0$ so that the angle becomes $n\pi t_0/T = \pi$. Since $\sin \pi = 0$, the amplitude of that particular harmonic is zero. This is called a spectrum null, or simply a null. As n is increased further, the harmonic amplitude increases, goes through a peak, and decreases until at $n = 2T/t_0$, when the angle is equal to 2π, there is again a spectrum null. This process of peaks of decreasing amplitude and nulls continues ad infinitum. A plot of Equation (3-20) for $t_0/T = 0.1$ is given in Figure 3-5(A). Just as for the spectrum of the squarewave train shown in Figure 3-2(C), each of the vertical lines in Figure 3-5(A) represents the amplitude of a sinusoid.

Except for the Gibbs phenomenon, we get back the original pulse train when these sinusoids are added in appropriate phase. It should again be emphasized that the spectrum analyzer does not show phase, so phase information was not shown in Figure 3-5. As expected, the 10th, 20th, 30th ... harmonics, corresponding to an angle equal to multiples of π, go to zero. If the ratio of pulse width to interpulse period (t_0/T) were other than 10 to 1, other harmonics than numbers 10 or 20 would go to zero, but this would still occur when the angle is equal to a multiple of π. Thus, the total angle $(n\pi t_0/T)$ rather than the harmonic number (n) is the fundamental parameter. Therefore, as we contemplate the effect on the spectrum of changes in the ratio t_0/T, the horizontal scale will remain in units of total angle, representing radian frequency rather than harmonic number.

Consider now the effect on Equation (3-20) of an increase in the interpulse spacing T. Suppose, for example, T is tripled so that $t_0/T = 1/30$ rather than $1/10$. Since all the harmonics are multiplied by the factor t_0/T, all amplitudes will decrease to one-third their original size. The basic shape of the spectral distribution will remain the same $(\sin x)/x$ shape. Nulls will again appear where x is a multiple of π, which happens when the harmonic number n is a multiple of T/t_0. As the ratio T/t_0 is increased, the number of components between nulls will also increase. In the above example, T/t_0 is increased to 30, so there are now thirty rather than ten harmonics between nulls. Or, looked at in another way, there are now three times as many signal components per unit frequency as there were before. Figure 3-5(B) is a plot of the spectrum of a rectangular pulse train with $T/t_0 = 30$.

As the ratio T/t_0 is increased by increasing the interpulse spacing (T), three important things happen. As $T \to \infty$, these are

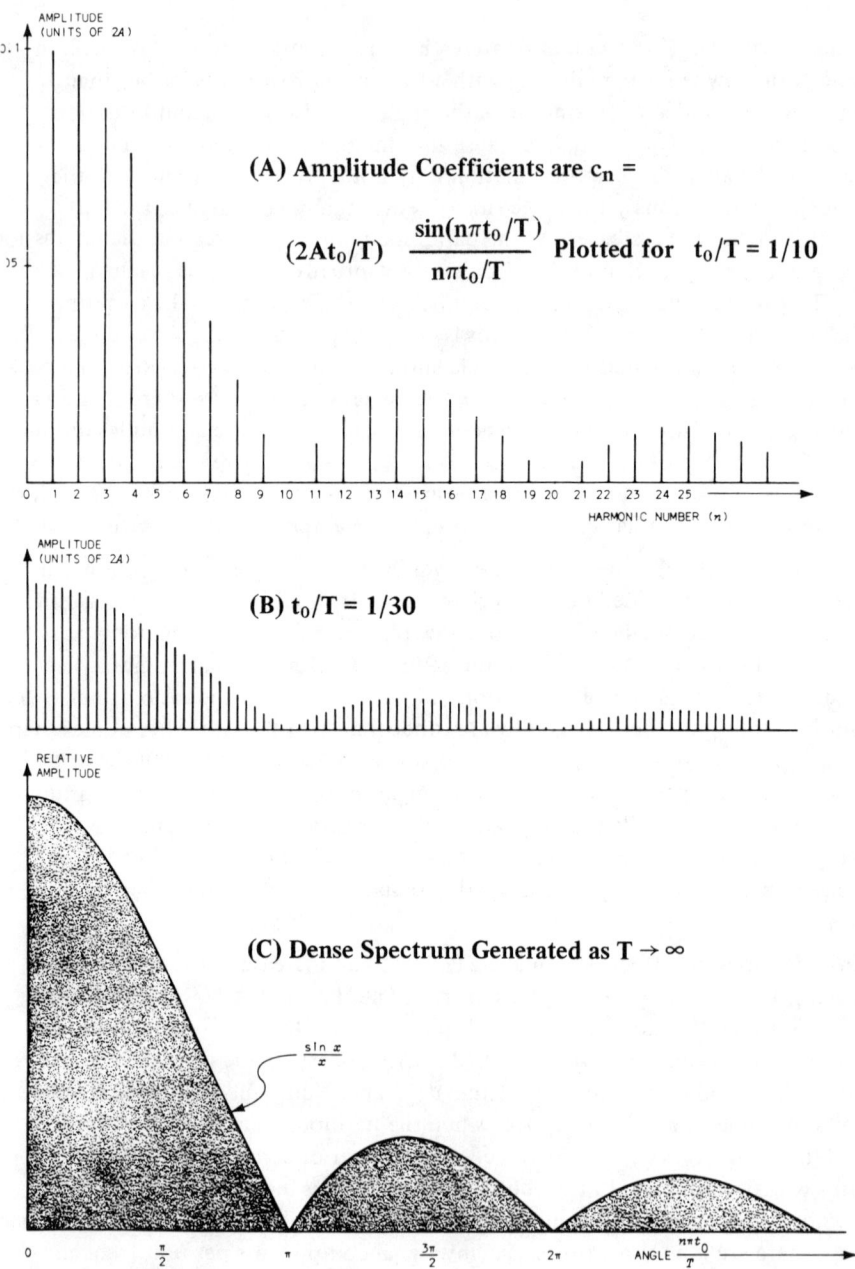

Figure 3-5 Frequency Distribution for Rectangular Pulse Train

Fourier Analysis 47

1. the amplitude of the individual components approaches zero,
2. the number of harmonics between nulls increases without bound, and
3. the shape of the curve is unchanged, remaining the same $(\sin x)/x$.

Ignore, for the moment, the apparent disappearance of everything as the amplitudes go to zero and concentrate on the meaning of the last two conclusions. As the number of harmonics increases without bound, a state is reached where it becomes impossible to distinguish between individual harmonics. Also, it would not make much sense to talk about individual harmonics as there are essentially an infinite number of these in any frequency interval, no matter how small. Yet, one parameter has remained unchanged as T was increased — the curve generated by the locus of the end points of the harmonic amplitudes. The shape of the curve is the previously discussed $(\sin x)/x$. Obviously the meaning of $x = n\pi t_0/T$ as $T \to \infty$ will have to be reinterpreted, otherwise x appears to approach zero. Ignoring this problem for the moment, we end with the graph in Figure 3-5(C).

Figure 3-5(C) is the frequency-domain representation of a dense, continuous spectrum. The spectrum is dense and continuous in the sense that, excepting the null points, no frequency can be found where there is no energy. Contrast this with the spectra of Figures 3-5(A) and 3-5(B), where there is energy only at specified frequencies, as indicated by the harmonics, and zero everywhere else. The usual procedure in establishing a description of dense spectra is to first establish the mathematical validity of the Fourier integral equations, which are then used in analyzing the spectra of single pulses. The procedure followed here is the opposite. First, using physical reasoning, an interpretation of the dense spectrum is developed that can then be used to justify the use of Fourier integrals. This method, while not mathematically rigorous, is helpful in establishing how a spectrum analyzer works. Those interested in a rigorous derivation are referred to the references.

There are two basic points that need to be considered when establishing a physical interpretation of Figure 3-5(C): how to handle the angle $n\pi t_0/T$ and what to do with the apparent disappearance of the spectrum since the ratio t_0/T seems to approach zero as T gets infinitely large. First the matter of the angle. It should be recognized that as the interpulse spacing T increases, so does the number of harmonics n occurring over any arbitrary frequency range. For example, the number of harmonics between two null points, which occur at angular differences of π, is $n = T/t_0$. Thus, as T goes toward infinity so does n and the ratio n/T remains constant.

In order to get rid of the bothersome infinities, it is, therefore, only necessary to treat the ratio n/T as a unit. This unit has the dimensions of inverse time interval, which is frequency; hence n/T is designated by the symbol f. It should be recognized that the use of the symbol f has greater significance than simple dimensional correctness. The frequency f is actually the frequency at which the harmonic n occurs. For example, suppose T = 1 ms, then the fundamental is at 1/T = 1 kHz, the second harmonic is at 2 kHz, the tenth harmonic at 10 kHz, etc., with the frequency of the nth harmonic at n/T. If, in addition, t_0 happened to be one-tenth the size of T, or 100 μs, then the tenth harmonic at 10 kHz would have zero amplitude, according to Figure 3-5(A). Note that the frequency of the tenth harmonic is not affected by t_0. Only n/T has to do with frequency, while t_0/T determines amplitude. Based on the above reasoning, the angle $n\pi t_0/T$ is replaced by $\pi f t_0$, thus eliminating all problems with infinite T.

The amplitude coefficient of Equation (3-20) is $2At_0/T$. Since A is pulse amplitude and t_0 is pulse width, the product At_0 is pulse area. The division by the interpulse period T is an averaging process, so that what is involved is the average pulse area. The factor 2 arises because theoretically the spectrum is symmetrical about the main lobe with center at n = 0. This point was discussed before in connection with the complex form of the Fourier series where for every coefficient of positive frequency d_n there is a corresponding conjugate of d_{-n}. Since, in practical spectrum analysis, negative frequencies have no meaning, the factors d_n and d_{-n} were combined to avoid confusion. For the rectangular pulse train d_{-n} and d_n are equal, which leads to an overall amplitude $C_n = 2d_n$. In any event, the conceptual difficulty is not with the factor 2 but with 1/T. The reason everything seems to go to zero is that Equation (3-20) deals with the amplitude of individual harmonics; but, as previously discussed, individual harmonics have no meaning when dealing with a continuous spectrum. This is because there are apparently an infinite number of these. Obviously, if individual harmonics contained a finite amount of energy, no matter how small, the total energy of all the harmonics would become infinite, and that is physically impossible. Instead of dealing with individual harmonics, it is necessary to deal with a spectral density or energy per unit bandwidth. This problem is analogous to that of the impulse function, where the parameters are essentially zero width, infinite amplitude, and finite area. Similarly, here the parameters are essentially zero amplitude, an infinite number of harmonics, and finite total energy. When using the spectral-density concept, the bothersome 1/T, actually frequency, is set equal to unity to represent a per-unit-bandwidth operation.

Based on the above, the discrete spectra components of Equation (3-20) are transformed to the continuous spectral-density distribution given in Equation (3-22),

$$F(\omega) = 2At_0 \frac{\sin(\pi t_0 f)}{\pi t_0 f} \qquad (3\text{-}22)$$

where $F(\omega)$ stands for a Fourier integral representation.

3.7. FOURIER INTEGRAL

The relationships for the complex form of the Fourier series are given in Equations (3-6) and (3-7) and reproduced below:

$$f(x) = \sum_{n=-\infty}^{n=+\infty} d_n \, \epsilon^{jnx} \tag{3-23}$$

$$d_n = \frac{1}{2\pi} \int_{-\pi}^{+\pi} f(x) \, \epsilon^{-jnx} \, dx \tag{3-24}$$

These equations are applicable when dealing with a discrete spectrum generated by a waveform having a finite period. This permits Equation (3-23) to be the sum of a discrete series of sinusoids, one for each n as n takes on all the positive and negative integer numbers. The series is infinite; there are an infinite number of integers, but the spectrum is not continuous, since all except very specific values of n are forbidden.

The finite period is clearly evident from Equation (3-24) where the limits of integration are $+\pi$ and $-\pi$. Here the period is assumed to be not only finite but specifically 2π. The equation can, of course, be modified for an arbitrary period T rather than 2π, as shown in Equation (3-12).

These equations have to be modified when dealing with isolated pulses, whose period is essentially infinite and which have continuous rather than discrete spectra. The continuous nature of the spectrum requires integration rather than summation, while the limits have to be extended to include a time function that never seem to end. The two integral equations replacing Equations (3-23) and (3-24) are called a Fourier transform pair. These are the direct transform:

$$F(\omega) = \int_{-\infty}^{+\infty} f(x) \, \epsilon^{-j\omega x} \, dx \tag{3-25}$$

and the inverse transform:

$$f(x) = \frac{1}{2\pi} \int_{-\infty}^{+\infty} F(\omega) \, \epsilon^{j\omega x} \, d\omega \tag{3-26}$$

Equations (3-25) and (3-26) are the complex notation versions of the Fourier integral. Although equivalent noncomplex notation equations, corresponding to similar equations for the Fourier series, can be developed, the complex notation is the easier to use.

Complex notation is helpful in Fourier transform useage because it leads to a pair of symmetrical equations. Thus, except for the factor $1/2\pi$, which comes from using the variable ($\omega = 2\pi f$) and the change of sign in the ($j\omega x$) exponents, the two equations are identical. This means that except for a change of scale, functions and their transforms are interchangeable. The fact that the frequency-domain representation of a rectangular pulse (direct transform) is (sin x)/x indicates that to get a rectangular spectral distribution one has to start with a (sin x)/x time-domain function.

As an example in using Fourier transforms, consider a rectangular pulse of width t_0 and amplitude A, such as one of the pulses shown in Figure 3-2(A). The time-domain function is f(t) = A, when $-(t_0/2) < t < +(t_0/2)$, and zero everywhere else. The direct Fourier transform equation is

$$F(\omega) = \int_{-\infty}^{+\infty} f(t)\, e^{-j\omega t}\, dt$$

Substituting for f(t) the value A between the limits of $t_0/2$ and $-(t_0/2)$:

$$F(\omega) = A \int_{-(t_0/2)}^{+(t_0/2)} e^{-j\omega t}\, dt$$

$$= A \frac{(-1)}{j\omega} e^{-j\omega t} \Big|_{-(t_0/2)}^{+(t_0/2)}$$

$$= A \frac{1}{j\omega} \left(e^{j\omega(t_0/2)} - e^{-j\omega(t_0/2)} \right)$$

Substituting Euler's identity as discussed in Chapter 2, the result is:

$$F(\omega) = A t_0 \frac{\sin \pi f t_0}{\pi f t_0} \tag{3-27}$$

where $\omega = 2\pi f$.

3.8. RECONCILING THEORY AND MEASUREMENT

Except for the multiplication factor of 2, Equation (3-27) is identical to Equation (3-22), which was developed through physical reasoning. As previously indicated, the factor 2 can be considered as arising from the elimination of energy at negative frequencies. However, there is more involved here, because the factor $1/2\pi$ in Equation (3-26) can be apportioned in several different ways without violating the self-consistency of the Fourier transform pair. Thus, the elimination of the factor 2 can be considered a normalization procedure where the Fourier integral yields the shape but not the absolute amplitude of the frequency spectrum. In order to determine the amplitude relationship between the time

Fourier Analysis

domain pulse and the measured frequency spectrum, it is necessary to consider the time response of a filter to a narrow pulse input. This is discussed in Chapters 5 and 9. The problem of discrete signals, for which Fourier series apply, is considered in the section on Waveform Analysis in Chapter 10. Equation (3-27) is plotted in Figure 3-6. Comparison of Figures 3-5(C) and 3-6 shows that the major difference between theory and practice is that in theory alternate lobes are of opposite phase, whereas in practice, no such distinction is made.

A table of some common transform pairs and properties of Fourier transforms appears at the end of this chapter. Of major significance is the center-frequency-shift property as a function of complex modulation. Basically, when a time function is multiplied by a carrier at frequency ω_0, the spectrum of that function is shifted by ω_0. Thus, for the rectangular pulse, whose spectral distribution is shown in Figure 3-6, the main lobe centered at $\pi f t_0 = 0$ moves from dc to ω_0. The frequency f in the angle $\pi f t_0$ becomes the difference frequency between the carrier at f_0 and the frequency of interest. This eliminates negative frequencies so long as $f_0 - f > 0$. Therefore, for a pulsed RF wave, the spectral distribution as shown on a spectrum analyzer is symmetrical about the main lobe as in Figure 3-6 rather than unidirectional as in Figure 3-5(C).

An interesting and useful characteristic of Fourier integral analysis is that the results are valid not only for single pulse or transient phenomena but for pulse trains as well. Note that the shape of the curve generated by connecting the end points of the harmonic amplitudes in Figures 3-5(A) and 3-5(B) is the same (sin x)/x as that of Figure 3-5(C). Therefore, all that is needed to reconstruct the harmonic amplitudes for a pulse train is a knowledge of how frequently to sample the continuous curve by taking the Fourier transform of the pulse.

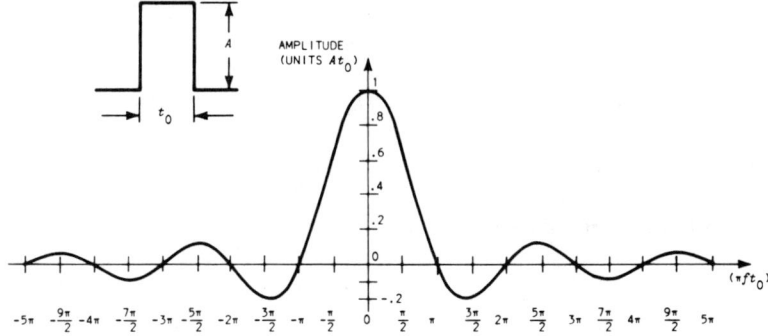

Figure 3-6 Fourier Transform of Rectangular Pulse

3.9. PROOFS AND TABLES

3.9.1. The Integral Equations for the Fourier Coefficients

The basic Fourier series given by Equation (3-1) is:

$$f(x) = \frac{a_0}{2} + (a_1 \cos x + b_1 \sin x) + (a_2 \cos 2x + b_2 \sin 2x) \cdots$$
$$+ (a_n \cos nx + b_n \sin nx) \tag{3-28}$$

To determine the three coefficients a_0, a_n, and b_n, it is necessary to evaluate the following three integrals:

$$\int_{-\pi}^{+\pi} f(x)\, dx$$

$$\int_{-\pi}^{+\pi} f(x) \cos nx\, dx \tag{3-29}$$

$$\int_{-\pi}^{+\pi} f(x) \sin nx\, dx$$

The evaluation of the integrals in Equation (3-29) is performed by substituting the series in Equation (3-28) for f(x) and integrating term by term. At first this appears to be an impossible task since Equation (3-28) is an infinite series. However, in each of the three integrals, all the terms except one yield zero. This stems from the following basic integral relationships for sinusoids:

$$\int_{-\pi}^{+\pi} \cos nx \cos mx\, dx = 0 \qquad m \neq n$$

$$\int_{-\pi}^{+\pi} \sin nx \sin mx\, dx = 0 \qquad m \neq n \tag{3-30}$$

$$\int_{-\pi}^{+\pi} \sin nx \cos mx\, dx = 0$$

The relationships in Equations (3-30) are simply an expression of the fact that sines and cosines form an orthogonal set of functions.

Fourier Analysis

Two special cases of the above are:

$$\int_{-\pi}^{+\pi} \cos mx \, dx = 0$$

(3-31)

$$\int_{-\pi}^{+\pi} \sin mx \, dx = 0$$

Using Equations (3-30) and (3-31), the integrals in Equation (3-29) are easily evaluated as follows:

$$\int_{-\pi}^{+\pi} f(x) \, dx = \int_{-\pi}^{+\pi} \frac{a_0}{2} \, dx + \int_{-\pi}^{+\pi} a_1 \cos x \, dx$$

$$+ \int_{-\pi}^{+\pi} b_1 \sin x \, dx$$

$$+ \ldots \int_{-\pi}^{+\pi} a_n \cos nx \, dx \qquad (3\text{-}32)$$

$$+ \int_{-\pi}^{+\pi} b_n \sin nx \, dx$$

But from Equations (3-30) and (3-31) all the terms except the first are zero, so

$$\int_{-\pi}^{+\pi} f(x) \, dx = \int_{-\pi}^{+\pi} \frac{a_0}{2} \, dx = \pi a_0 \qquad (3\text{-}33)$$

which is the same as Equation (3-3) given as

$$a_0 = \frac{1}{\pi} \int_{-\pi}^{+\pi} f(x) \, dx$$

To determine a_n we evaluate

$$\int_{-\pi}^{+\pi} f(x) \cos nx \, dx = \int_{-\pi}^{+\pi} \frac{a_0}{2} \cos nx \, dx$$

$$+ \int_{-\pi}^{+\pi} a_1 \cos x \cos nx \, dx$$

$$+ \int_{-\pi}^{+\pi} b_1 \sin x \cos nx \, dx$$

$$+ \quad \ldots \int_{-\pi}^{+\pi} a_n \cos nx \cos nx \, dx$$

$$+ \int_{-\pi}^{+\pi} b_n \sin nx \cos nx \, dx \qquad (3\text{-}34)$$

Based on Equations (3-30) and (3-31), all the terms in Equation (3-34) are zero except one, which leaves

$$\int_{-\pi}^{+\pi} f(x) \cos nx \, dx = \int_{-\pi}^{+\pi} a_n \cos nx \cos nx \, dx$$

which when evaluated leads to the result

$$\int_{-\pi}^{+\pi} f(x) \cos nx \, dx = \int_{-\pi}^{+\pi} a_n \cos^2 nx \, dx = \pi a_n \qquad (3\text{-}35)$$

Equation (3-35) is identical with that of Equation (3-3):

$$a_n = \frac{1}{\pi} \int_{-\pi}^{+\pi} f(x) \cos nx \, dx$$

Similar reasoning leads to the result that

$$b_n = \frac{1}{\pi} \int_{-\pi}^{+\pi} f(x) \sin nx \, dx \qquad (3\text{-}36)$$

Equations (3-33), (3-35) and (3-36) are used in evaluating the coefficients in the Fourier series expansion.

3.8.2. The Impulse Function

The time-domain unit impulse, designated delta of t, $\delta(t)$, is a function having an infinitely narrow pulse width and unity area, so that its amplitude approaches infinity. Mathematically, a unit impulse at time t = 0 has the property that:

Fourier Analysis

$$\delta(t) = 0, \quad t \neq 0$$
$$\delta(t) \to \infty, \quad t = 0$$
$$\int_{-\epsilon}^{+\epsilon} \delta(t)\,dt = 1 \qquad (3\text{-}37)$$

where epsilon is an arbitrarily small time interval.

The area of the impulse is called its strength, so that a unit impulse has a strength of one. Naturally, it is not mandatory that all impulses have unit strength; any strength at all is possible. The unit impulse is, however, a convenient quantity to manipulate, so that other impulses are defined in terms of the unit impulse.

The impulse is not a practically realizable function because real circuits cannot generate infinitely narrow arbitrarily large pulses. From a more fundamental point of view, the impulse is not realizable because the energy content of an impulse is infinite. This is because energy is proportional to the square of the impulse function. Thus the area of an impulse is

$$\int_{-\infty}^{+\infty} \delta(t)\,dt$$

and is finite, but the energy is given by

$$\int_{-\infty}^{+\infty} [\delta(t)]^2\,dt$$

which is infinite. This matter of infinite energy becomes clearer when the impulse is considered in the frequency domain. Thus, taking the Fourier transform of the unit impulse

$$F(\omega) = \int_{-\infty}^{+\infty} \delta(t)\, e^{-j\omega t}\,dt = 1$$

The fact that the spectral distribution of an impulse is a constant means that the impulse has a constant energy per unit bandwidth at all frequencies. Hence, going to higher and higher frequencies causes the energy to increase without bound. What happens in real life is that the spectral density starts falling off at some arbitrarily high frequency so that the total energy remains finite.

Since mathematical impulses do not exist in nature, it is of interest to investigate various approximations. The impulse can be approximated by any pulse such that the area remains constant as the width is decreased. A rectangular pulse of width τ and height $1/\tau$ has a constant unity area for all τ and is therefore an acceptable representation of the unit impulse. The exponential function $(1/\alpha)e^{-(t/\alpha)}$, shown graphically in Figure 3-7, is frequently used to represent an impulse. The area under this curve, as α goes to zero, is

$$\text{Area} = \int_0^\infty f(t)\,dt = \frac{1}{\alpha}\int_0^\infty \epsilon^{-(t/\alpha)}\,dt = -\epsilon^{-(t/\alpha)}\Big|_0^\infty$$

Determining $\lim_{\alpha \to 0} \int_0^\infty f(t)\,dt$:

$$\lim_{\alpha \to 0} -\epsilon^{-(t/\alpha)}\Big|_0^\infty = \lim_{\substack{\alpha \to 0 \\ t \to 0}} \epsilon^{-(t/\alpha)} - \lim_{\substack{\alpha \to 0 \\ t \to \infty}} \epsilon^{-(t/\alpha)} = 1 \tag{3-38}$$

The exponential curve is, therefore, a good beginning shape for the unit impulse. The shape of the starting pulse really doesn't matter as long as the area remains constant as the pulse width is reduced.

As previously indicated, the pulse width can never be reduced to zero. However, when the pulse width is reduced to the point where the spectral distribution is constant over the frequency range of the circuits used, for all practical purposes, an impulse has been generated.

The flat frequency distribution of the impulse is connected with a property of the Fourier transform pair, which is sometimes called reciprocal spreading. When one member of the transform pair is made narrower, the other spreads out and vice versa. Thus, for the case of the rectangular pulse, the first null occurs when the angle $\pi f t_0 = \pi$, which happens at the frequency $f = 1/t_0$. As the pulse width t_0 is made narrower, the frequency width occupied by the main lobe, as is illustrated in Figure 3-6, gets wider and wider until, as the pulse width goes to zero, the peak of the main lobe is spread out over all frequencies.

Conversely, as the pulse width t_0 is made wider, all of the energy gets more and more concentrated at the center frequency of the main lobe until, as the pulse width approaches infinity, the complete frequency distribution gets concentrated in an infinitesimally thin frequency band. This is, of course, a frequency impulse as discussed in connection with the Fourier transform of an infinitely long sinusoid. The frequency impulse is quite similar in its properties to the time impulse. All of the mathematics are almost identical except for the change of variable from t to f. This should not be surprising in view of the symmetry of the Fourier transform pair. Like its time domain counterpart, true frequency domain impulses do not exist in nature, since an infinitely long sinusoid is obviously impossible to generate. However, as long as the sinusoid exists for a long time compared to the time constants of the circuits involved, it can be treated as if it were of infinite duration.

3.9.3. Properties of Fourier Transforms

A knowledge of some of the properties of Fourier transforms can be very help-

Fourier Analysis

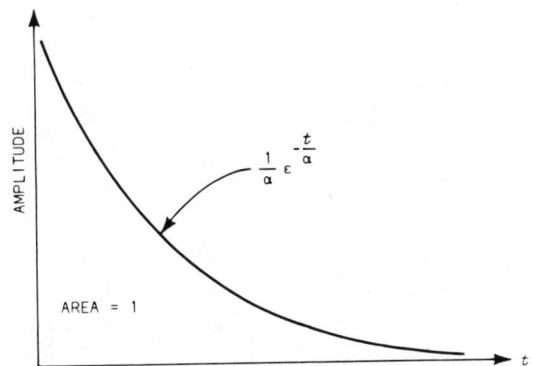

Figure 3-7 Exponential Prototype of Impulse

ful in spectrum analysis. For example, the fact that superposition holds for Fourier analysis can be helpful when interpreting complex spectra. Table 3-1 is a list of the more significant Fourier transform properties as found in spectrum analysis. No proofs are included. Some of these properties, such as convolution, have well-established names, while others may be found under different names in different tables. Table 3-2 gives both graphical and mathematical relationships for time-domain to frequency-domain conversion.

3.10. EXAMPLES

3.10.1. Using One Spectrum to Derive Another

Frequently, the spectral distribution of one waveform can be obtained by using the spectral distribution of another waveform. As an example, consider the half-cycle cosine pulse and the pulsed RF rectangular pulse, numbers 3 and 5 respectively in Table 3-2.

The spectrum for a rectangular RF pulse of unity amplitude is from number 5 in Table 3-2:

$$F(f) = \frac{t_0}{2} \left(\frac{\sin \frac{1}{2}(\omega - \omega_0)t_0}{\frac{1}{2}(\omega - \omega_0)t_0} + \frac{\sin \frac{1}{2}(\omega + \omega_0)t_0}{\frac{1}{2}(\omega + \omega_0)t_0} \right) \qquad (3\text{-}39)$$

The product $\omega_0 t_0$ is the radian angle through which the carrier at ω_0 advances during the time t_0. For a half-cycle pulse, $\omega_0 t_0 = \pi$, since a full cycle is 2π radians. Substituting $\pi = \omega_0 t_0$ and cancelling terms results in

Table 3-1
Time and Frequency Domain Relationships

NAME	TIME DOMAIN	FREQUENCY DOMAIN	COMMENTS
1) DIRECT TRANSFORM	$f(t)$	$F(\omega) = \int_{-\infty}^{+\infty} f(t) e^{-j\omega t} dt$	Fourier transform pair in complex notation.
2) INVERSE TRANSFORM	$f(t) = \frac{1}{2\pi}\int_{-\infty}^{+\infty} F(\omega) e^{j\omega t} d\omega$	$F(\omega)$	
3) SUPERPOSITION	$f_1(t) + f_2(t)$	$F_1(\omega) + F_2(\omega)$	The Fourier transform of the sum of several time-domain functions is equal to the sum of the individual transforms. In this respect superposition is equivalent to addition.
4) DUALITY OF TRANSFORM PAIR	$f(t)$	$F(\omega)$	The transform pair differ from each other only in a scale change of 2π and in the sign of an exponent. Thus, if a time-domain shape (e.g., rectangle) gives a certain frequency shape, e.g., $(\sin x)/x$, then except for a scale change and sign inversion, a time shape like the frequency shape $[(\sin x)/x$ in time] will result in a frequency shape like the time shape (rectangular pulse in frequency domain).
	$F(t)$	$2\pi f(-\omega)$	
	$\frac{1}{2\pi}F(-t)$	$f(\omega)$	
5) CHANGE OF SCALE (RECIPROCAL SPREADING)	$f(at)$	$\frac{1}{a}F\left(\frac{\omega}{a}\right)$	This is related to the phenomenon which is sometimes referred to as reciprocal spreading. As one member of a Fourier transform pair narrows, the other spreads out. Furthermore, as the function spreads out the amplitude is reduced. This is a necessary consequence of the requirement that the total energy be the same whether the function is considered in time or frequency. Reciprocal spreading seems to be a universal property that applies to many reciprocal parameters. As the precision with which one parameter can be known is increased, the precision with which the other can be known is decreased. The best known expression of this type is the Heisenberg uncertainty principle in quantum mechanics which states that the product of the measurement accuracy of two reciprocal parameters, such as position and momentum, cannot be less than Planck's constant. Mathematically: $\Delta p \Delta q = \hbar$ Heisenberg principle. This is similar to: $\Delta f \Delta t = 1$ Fourier theory.
6) EQUAL ENERGY OR PARSEVAL'S THEOREM	$\int_{-\infty}^{+\infty}[f(t)]^2 dt$	$\frac{1}{2\pi}\int_{-\infty}^{+\infty}\lvert F(\omega)\rvert^2 d\omega$	The energy computed by taking the integral of the square of the time-domain function or the integral of the square of absolute value of the frequency-domain function is the same. In other words, the time- and frequency-domain representations are equivalent as far as energy is concerned.

Fourier Analysis

7) TIME SHIFT	$f(t-\tau)$	$e^{-j\omega\tau} F(\omega)$	A change in position by $-\tau$ in the time domain introduces a phase shift of $\phi = -\omega\tau$ in the frequency domain. Conversely, introducing a phase shift in the frequency domain is equivalent to a delay in the time domain.
8) FREQUENCY SHIFT	$e^{j\omega_0 t} f(t)$	$F(\omega-\omega_0)$	This is the inverse or dual of the time shift. The time operation is that of multiplication by another frequency, which in engineering is ordinarily called modulation. Thus, a shift in the frequency of a carrier that is pulsed on and off introduces a corresponding shift in the spectrum of the pulse.
9) CONVOLUTION	$f_1(t) f_2(t)$	$\frac{1}{2\pi}\int_{-\infty}^{+\infty} F_1(p) F_2(\omega-p) \, dp$	The spectrum of the product of two time-domain functions is the convolution of the individual spectra, where convolution is defined by the integral. The meaning of convolution is easier to understand by considering the inverse operation in the time.
	$\int_{-\infty}^{+\infty} f_1(\tau) f_2(t-\tau) \, d\tau$	$F_1(\omega) F_2(\omega)$	Convolution occurs when a signal is developed by sliding two functions past each other. Any scanning function, such as a filter sliding relative to a signal in spectrum analysis, involves convolution. The convolution of a function with an impulse results in the same shape as the original function. Hence, in spectrum analysis, if the scanning filter is assumed to be sufficiently narrow, the result of sliding the signal past the filter is the spectral characteristic of the signal.
10) DIFFERENTIATION	$\frac{d}{dt} f(t)$	$j\omega F(\omega)$	If the spectrum of $f(t)$ is given by $F(\omega)$, then the spectrum of the derivative of $f(t)$ is just $j\omega$ times that of $f(t)$. In other words, differentiation in the time domain means multiplication by $j\omega$ in the frequency domain.
11) INTEGRATION	$\int_{-\infty}^{t} f(t) \, dt$	$\frac{1}{j\omega} F(\omega)$	Integration is essentially the inverse of differentiation. Integration in the time domain means multiplication by $\frac{1}{j\omega}$ in the frequency domain.

Table 3-2
Fourier Transforms
(Format Courtesy of Hewlett Packard Company)

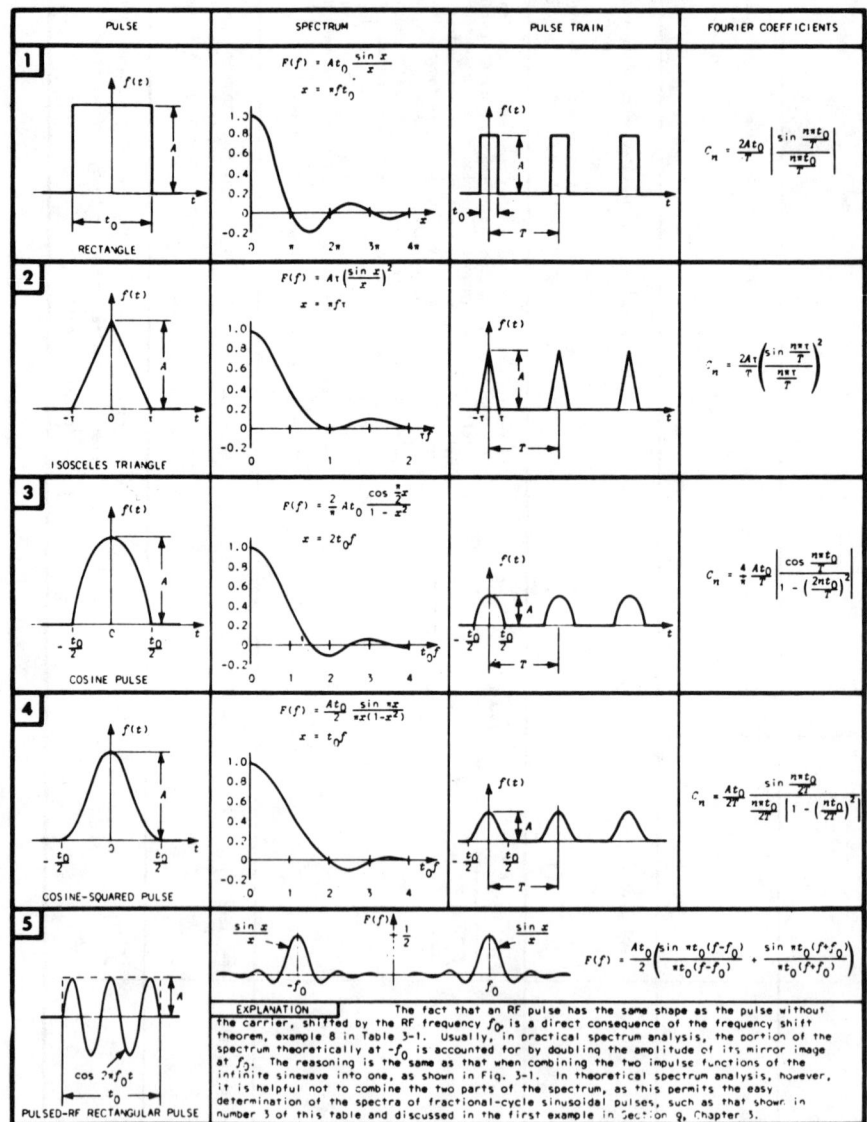

Fourier Analysis

$$F(f) = \frac{\sin \frac{1}{2}(\omega t_0 - \pi)}{\omega - \omega_0} + \frac{\sin \frac{1}{2}(\omega t_0 + \pi)}{\omega + \omega_0} \tag{3-40}$$

The factor t_0 is eliminated by substituting $t_0 = \pi/\omega_0$, which leads to

$$F(f) = \frac{1}{\omega - \omega_0} \sin\left(\frac{\pi\omega}{2\omega_0} - \frac{\pi}{2}\right) + \frac{1}{\omega + \omega_0} \sin\left(\frac{\pi\omega}{2\omega_0} + \frac{\pi}{2}\right) \tag{3-41}$$

Equation (3-37) can be further simplified by using the basic trigonometric identities:

$$\sin(A + B) = \sin A \cos B + \cos A \sin B$$

$$\sin(A - B) = \sin A \cos B - \cos A \sin B \tag{3-42}$$

Using Equations (3-38) in (3-37) we have:

$$F(f) = \frac{1}{\omega - \omega_0}\left(\sin \frac{\pi\omega}{2\omega_0} \cos \frac{\pi}{2} - \cos \frac{\pi\omega}{2\omega_0} \sin \frac{\pi}{2}\right)$$

$$+ \frac{1}{\omega + \omega_0}\left(\sin \frac{\pi\omega}{2\omega_0} \cos \frac{\pi}{2} + \cos \frac{\pi\omega}{2\omega_0} \sin \frac{\pi}{2}\right) \tag{3-43}$$

But, $\sin(\pi/2) = 1$ and $\cos(\pi/2) = 0$, hence Equation (3-43) reduces to

$$F(f) = \left(-\frac{1}{\omega - \omega_0} \cos \frac{\pi\omega}{2\omega_0}\right) + \left(\frac{1}{\omega + \omega_0} \cos \frac{\pi\omega}{2\omega_0}\right) \tag{3-44}$$

Combining terms we have

$$F(f) = \left(\frac{1}{\omega + \omega_0} - \frac{1}{\omega - \omega_0}\right) \cos \frac{\pi\omega}{2\omega_0} \tag{3-45}$$

Combining the terms in the parenthesis by means of the common denominator, $\omega^2 - \omega_0^2$, results in

$$F(f) = \frac{-2\omega_0}{\omega^2 - \omega_0^2} \cos \frac{\pi\omega}{2\omega_0} \tag{3-46}$$

Normalizing with respect to ω_0 by letting $\omega/\omega_0 = x$ and substituting π/t_0 for ω_0:

$$F(f) = \frac{2t_0}{\pi} \frac{1}{1 - x^2} \cos \frac{\pi}{2} x \tag{3-47}$$

This is the same expression as that given in number 3 in Table 3-2 with the amplitude A set equal to unity.

While it took a bit of algebra and trigonometry, this exercise demonstrates the fact that by knowing a few key transforms, it is possible to obtain others without going through a solution of the (sometimes difficult) integral equations.

3.10.2. The Sine Integral, Si(X)

When dealing with continuous spectra, it is inappropriate to operate in terms of individual harmonics. The proper way to consider the spectral distribution is in terms of spectral density, which is a per-unit frequency difference, or bandwidth, quantity. Thus, in actual measurements, the wider the bandwidth of the measuring apparatus the more energy should be intercepted and the greater should be the output indication. But, most spectra do not have a flat frequency spectrum, so the output depends on where the apparatus passband intercepts the spectral distribution. For the rectangular pulse, for example, there would be considerable output when the measuring filter is tuned to the center of the main lobe and very little output when the filter frequency is at a null point. This variation in spectral density is of course determined by the area under any small portion of the spectral distribution curve, such as in Figure 3-6.

To obtain an area, it is necessary to integrate. This leads to the importance of the *sine integral* Si(x), where

$$\text{Si}(x) = \int_0^x \frac{\sin x}{x} \, dx \tag{3-48}$$

Since no area is intercepted at zero bandwidth at $x = 0$, $\text{Si}(x) = 0$. This is in agreement with the previous reasoning that led to the conclusion that the amplitude of an individual spectral line, which has zero frequency width, is zero. As x increases toward the first zero crossing at π, Si(x) keeps increasing until, at $x = \pi$, Si(x) = 1.85. As x goes greater than π, Si(x) starts decreasing because the curve (sin x)/x is now negative. At $x = 2\pi$, Si(x) starts to increase again, oscillating back and forth every time x increases by π. In order to treat negative-going areas equally with positive-going areas, one needs either to take the integral of $[(\sin x)/x]^2$ or, as is more common, to determine the difference between two sine integrals. Thus, if dealing with an instrument bandwidth ΔF and wishing to know the output around frequency f, the value of $\text{Si}(x_2) - \text{Si}(x_1)$ must be determined where

$$x_2 = \pi t_0 \left(f + \frac{\Delta f}{2} \right)$$

$$x_1 = \pi t_0 \left(f - \frac{\Delta f}{2} \right)$$

The availability of sine integral tables can, therefore, be quite useful. Fortunately, this integral occurs in many communication problems so that tables are readily available. Figure 3-8 is a plot of this integral.

Fourier Analysis

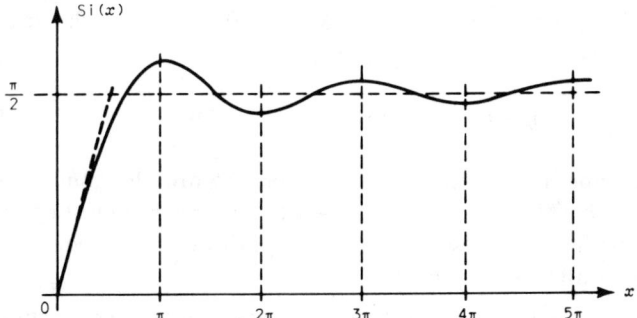

Figure 3-8 A Plot of the Sine Integral of X

3.10.3. Rectangular Pulse Analysis

Given a rectangular pulse, determine for its spectrum the position and amplitude of the first side lobe relative to the main lobe. The spectral distribution is given by the formula $F(f) = (\sin x)/x$, where $x = \pi f t_0$. A plot is given in Figure 3-6. To determine the position of the various maxima, the expression must be differentiated and set equal to zero. Thus:

$$F(f) = \frac{\sin x}{x}$$

$$\frac{d}{dx} F(f) = \frac{x \cos x - \sin x}{x^2} = 0$$

which means that maxima occur when

$$x = \tan x \qquad (3\text{-}49)$$

Observing the graph of $(\sin x)/x$ (Figure 3-6), the maximum of the first side lobe occurs in the vicinity of $x = \pi/2$. By trial-and-error comparison of the value of $\tan x$ with x, the actual angle is found to be close to

$$x = 4.5 \text{ radians} \qquad (3\text{-}50)$$

which is 1.43π rather than the 1.5π estimated from Figure 3-6. The relative amplitude is determined by substituting the appropriate value of x into the $(\sin x)/x$ equation. Thus, for the peak of the main lobe, $x = 0$. At small angles the sine of an angle is equal to the angle, so that

$$\lim_{x \to 0} \frac{\sin x}{x} = 1$$

The amplitude of the first side lobe is (sin 4.5)/4.5 = 0.2172. The relative amplitude between the main lobe and first side lobe is

$$\frac{1}{0.2175} = 4.6, \text{ or } 20 \log 4.6 = 13.2 \text{ dB} \tag{3-51}$$

Very often this number is approximated as 13.4 or 13.5 dB. This comes from the approximation that the peak occurs at x = 1.5π radians. For most applications, this is well within the measurement accuracy. However, in precision measurements, 13.2 dB should be used.*

If the pulse width is t_0 = 1 μs, what is the frequency width of the spectrum lobes? The spectrum nulls occur at a spacing of x = π. Since x = $\pi f t_0$, this happens at frequency multiples of f = 1/t_0. Hence, the lobe width is 1/1 μs = 1 MHz.

What if the pulse width is increased to 10 μs? The lobe spacing then becomes 1/10 μs = 100 kHz. This is an example of reciprocal spreading, where, as the pulse width gets wider, the spectra width gets narrower.

What happens if the pulse consists of a 1 μs burst of a 1 GHz sinusoid? By the frequency shift theorem, number 8 in Table 3-1, the center frequency of the main lobe is shifted to 1 GHz. The spectral shape remains a (sin x)/x as before.

What are the characteristics of a 1 μs burst of a 500 kHz sinusoid? The first tendency is to say that the result is a (sin x)/x centered at 500 kHz. However, 1 μs is not sufficient time to establish the pulse shape of the gating oscillator when gating a 500 kHz carrier. As a matter of fact, we pass only a half cycle of 500 kHz during 1 μs. The result is the cosine pulse discussed previously in Example (A) and transform number 3 in Table 3-2.

How is the spectrum of a burst of a 1 GHz sinusoid modified if the sinusoid does not turn off completely, as shown in Figure 3-9?

If the 1 μs pulse width is a substantial part of the waveform cycle, we need to consider Figure 3-9(A) as composed of two pulses, one short and large in amplitude and the other long and small in amplitude or as a continuous sinewave and a pulsed sinewave riding on top of it. If, as is usually the case, the burst occupies only a small part of the waveform period and is very much larger in amplitude

---

*Any desired degree of accuracy can be obtained by a program of successive approximations. Thus, the first sidelobe peak, computed to six decimal places, occurs at 4.493409 radians and is 13.26159 dB below the mainlobe peak. Numbers for the first five sidelobes follow:

Sidelobe Number	1	2	3	4	5
Angle Radians	4.493	7.725	10.904	14.066	17.220
dB Down	13.261	17.830	20.788	22.985	24.735

Fourier Analysis

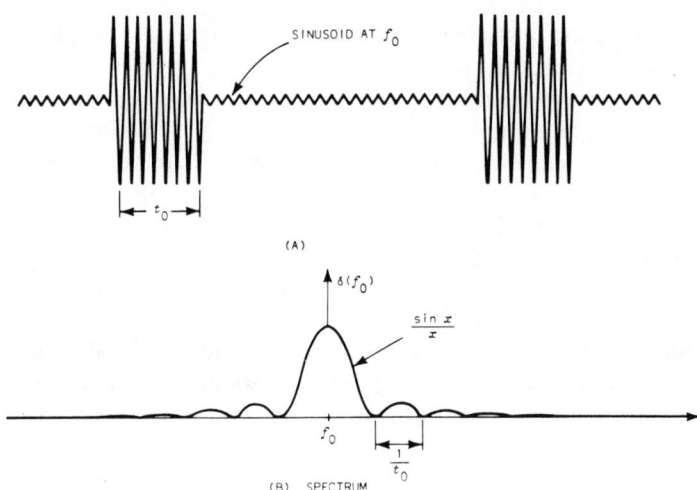

Figure 3-9 Pulsed RF with Poor On/Off Ratio

than the remaining sinusoid, the approximation of a full amplitude pulse co-existing with an uninterrupted sinusoid can be used. In either case, superposition applies, so that the waveform can be considered to be composed of two components. In the prevalent case of a large narrow pulse, the spectrum consists of the superposition of the (sin x)/x for the pulse and an impulse for the sinusoid, as shown in Figure 3-9(B).

3.11. EXERCISES

3-1. Determine the Fourier series for the waveform of Figure 2-2(C).
3-2. Verify the equality given by Equation (3-8).
3-3.
  a. Show that the Fourier coefficients for the sawtooth waveform shown in Figure 3-10 is: $C_n = A/\pi n$

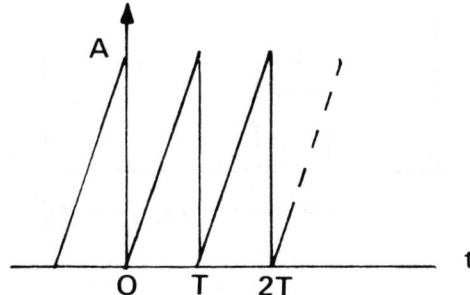

Figure 3-10 Problem 3-3

b. Compute the first five terms (including the dc term, if any) for the Fourier series of the sawtooth waveform.

3-4. The Fourier series representation of a squarewave is given by Equation (3-15). Note the absence of even harmonics. Which, if any, of pulse trains 1, 2, and 3 shown in Table 3-2 have only odd harmonics when duty factor is at 50%?

3-5. Given the two pulse trains shown below, do the Fourier series differ? If yes, how?

3-6. Show that the Fourier transform for a triangular pulse such as No. 2 in Table 3-2 is a $(\sin^2 X/X^2)$.

3-7. Show that for the $\sin X/X$ spectral distribution, the peak of the third sidelobe occurs at $X \cong 11$ radians and is about 20.8 dB down from the mainlobe.

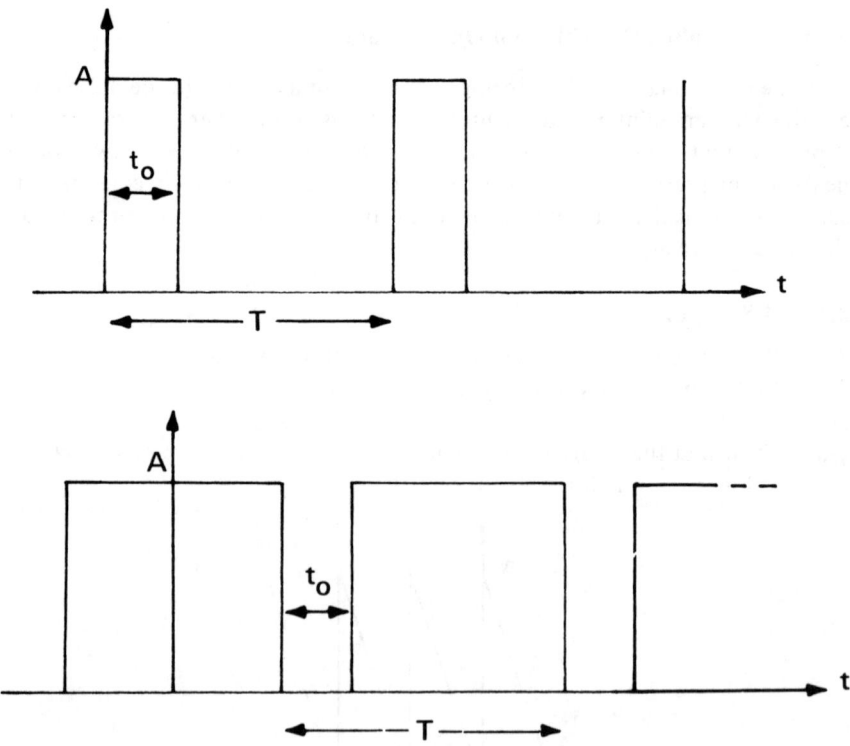

Figure 3-11

Fourier Analysis

3-8. The peak of the first sidelobe is about 13.2 dB below the peak of the mainlobe for a rectangular pulse. What is the equivalent number for an isosceles triangle pulse?

3-9. The Fourier series for a particular waveform is given by $f(X) = 3 \sin X - 2 \sin 2X + \sin 4X + 2 \cos X + 2 \cos 2X - \cos 3X$. What are the combined amplitudes (C_n) that would be shown by a spectrum analyzer?

3-10. Given a rectangular pulse train of pulse amplitude 100 and duty factor $(t_0/T) = 1/20$,
   a. What are the amplitudes of the 1st, 20th, 30th and 80th harmonics?
   b. What is the dense spectrum (Fourier integral) representation for a single pulse?
   c. What are the peak amplitudes for the first, second and third sidelobes?
   d. Assuming that $t_0 = 1/200$, what is the area of the mainlobe for the dense spectrum representation?

Chapter 4
Modulation Theory

4.1. INTRODUCTION

In electronic communications the message is usually not in a form suitable for transmission over the medium intervening between transmitter and receiver. The process whereby the original message is modified into an information-bearing transmittable signal is called modulation. Modulation theory is a vast subject that properly belongs as part of information theory. It is certainly not the intent, nor within the scope, of this volume to treat a subject of such complexity. Fortunately, the two forms of modulation (AM and FM), most often described in the frequency domain by measurement with spectrum analyzers, are also the most amenable to relatively simple mathematical analysis. Such important topics as pulse-code modulation (PCM) or time or frequency multiplexing will not be considered. Within the information-theory meaning of the word modulation, only amplitude modulation (AM) and angle modulation in the form of frequency modulation (FM) will be considered.

To discuss modulation, it is necessary to state three definitions:

1. Carrier: the wave to which modulation is applied.
2. Modulating wave: the signal which contains the original message and is used to control some parameter of the carrier.
3. Modulated wave (modulated carrier): the final result of the modulation process whereby the original message is modified into an information-bearing signal, that is sent by the transmitter to the receiver.

None of these three waves, or signals, need be sinusoidal. For example, in pulse modulation, the carrier consists of a train of pulses, some parameter of which, such as pulse height or pulse position, is controlled by the modulating wave. The modulating wave, possibly generated by speech, is certainly far from sinusoidal. Finally the modulated wave, consisting of some complex combination of the two above, can be quite complicated. Nevertheless, the analysis that follows is based on sinusoidal waveforms. This is justified on the basis that in AM and FM the carrier is a sinusoid and that any modulated wave can be broken into an equivalent series of sinusoids by means of Fourier analysis. The ultimate reason for doing things this way is, of course, the tremendous simplification in the analysis.

A sinusoid, A sin θ, has two basic parameters that can be varied: the amplitude A and the angle θ. Let us begin by analyzing the effect of a changing amplitude.

4.2. AMPLITUDE MODULATION

The carrier in amplitude modulation (AM) is usually a sinusoid of the form

$$A \sin(2\pi Ft + \alpha) \tag{4-1}$$

where A is the carrier amplitude, F is the carrier frequency, and α is the initial phase or just phase. Assume that the information desired to be transmitted is also sinusoidal in nature and represented mathematically by the modulating wave

$$B \cos 2\pi ft \tag{4-2}$$

What is meant by AM is that the amplitude of the carrier is made to vary in proportion to the modulated wave of the form:

$$a = A(1 + m \cos 2\pi ft) \sin(2\pi Ft + \alpha) \tag{4-3}$$

where m is called the degree of modulation, or 100m is the percentage modulation, f is the modulation frequency, and a is the instantaneous amplitude. Usually the word amplitude refers to a constant such as A in Equation (4-1). A more correct name for a might be instantaneous value. However, a is an amplitude in the sense that its square is proportional to instantaneous power. Figure 4-1 is a graphical representation of Equation (4-3).

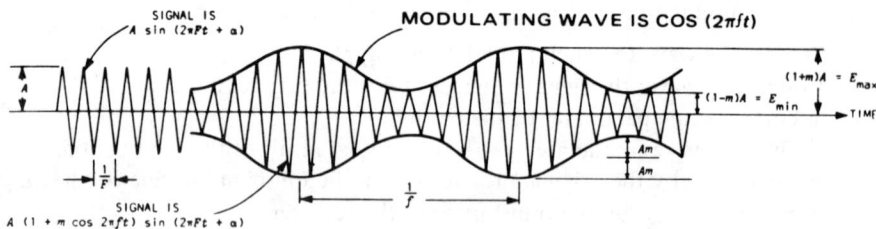

Figure 4-1 Time Domain Appearance of Amplitude Modulation

In order to obtain the frequency domain representation of an amplitude-modulated wave, it is necessary to disassociate the complex Expression (4-3) into a sum of individual sinusoids. This is easily accomplished with the help of the trigonometric identity

$$\sin A \cos B = \frac{1}{2}\left[\sin(A+B) + \sin(A-B)\right] \tag{4-4}$$

Modulation Theory

Letting $2\pi Ft + \alpha = A$ and $2\pi ft = B$ and substituting Expression (4-4) into Expression (4-3):

$$a = A\left[\{1 + m\cos(2\pi Ft)\}\sin(2\pi ft + \alpha)\right]$$

$$a = \underbrace{A\sin(2\pi Ft + \alpha)}_{\text{carrier}} + \underbrace{\frac{Am}{2}\sin\left[2\pi(F+f)t + \alpha\right]}_{\text{upper sideband}} \quad (4\text{-}5)$$

$$+ \underbrace{\frac{Am}{2}\sin\left[2\pi(F-f)t + \alpha\right]}_{\text{lower sideband}}$$

From Equation (4-5) it will be observed that an AM wave can be considered as consisting of the original carrier and two new components called sidebands. The sidebands are spaced on either side of the carrier with a frequency spacing equal to the modulating frequency f. The amplitude of the sidebands, relative to that of the carrier, is equal to half the percentage modulation, m/2. The frequency domain representation of an AM wave is shown in Figure 4-2.

From Equation (4-5), it is apparent that the carrier component of the AM spectrum is independent of the degree of modulation. Hence, an amplitude-modulated wave always contains more energy than the unmodulated carrier. At 100% modulation, the sidebands are half as large as the carrier. This is the maximum relative amplitude that the sidebands can attain without overmodulation. Since energy is proportional to voltage amplitude squared, it follows that at 100% modulation each sideband contains one-quarter as much energy as the carrier. Thus, at maximum modulation, the AM wave contains 50% more energy than the unmodulated carrier.

What is of interest in determining amplitude modulation is the carrier frequency F, the modulating frequency f, and the degree or percent modulation m. The frequencies are usually known beforehand so the degree of modulation is the most frequent measurement. All three of the above parameters can theoretically be determined by use of either a time domain oscilloscope or a frequency domain spectrum analyzer.

In the time domain, illustrated in Figure 4-1, the frequencies F and f are easily determined as the inverse of two simple time measurements, while the degree of modulation m is computed from a knowledge of the peak and null waveform amplitudes. Thus, the peak waveform amplitude is $E_{max} = (1+m)A$, while the null amplitude is $E_{min} = (1-m)A$. Calling the ratio of these some constant, K:

$$K = \frac{E_{max}}{E_{min}} = \frac{(1+m)A}{(1-m)A} \quad m = \frac{K-1}{K+1} = \frac{E_{max} - E_{min}}{E_{max} + E_{min}} \quad (4\text{-}6)$$

In Figure 4-1, K = 3, hence m = (3 − 1)/(3 + 1) = 0.5, or there is 50% of amplitude modulation.

While all the parameters can be obtained from time domain measurements in theory, there are practical difficulties. Problems with time domain measurements are:

1. Too high a carrier frequency for time domain measurement. Sampling oscilloscopes have greatly reduced this problem.
2. Too small an amplitude level to be observed on an oscilloscope.
3. The complicated nature of the signal when more than one modulating frequency is involved.
4. Difficulty in determining the ratio of E_{max}/E_{min} at low percentages of modulation.

To alleviate problems such as those above, it is customary to make AM measurements by means of the frequency domain spectrum analyzer. Here the carrier frequency F is obtained from the calibrated RF center frequency dial, the modulation frequency f is determined by measuring the frequency difference between the carrier and the sidebands, while the degree of modulation is determined by measuring the relative amplitude between carrier and sidebands and computing from

$$\% \text{ modulation} = m \cdot 100 = \frac{2 \cdot A_{sideband}}{A_{carrier}} \cdot 100 \tag{4-7}$$

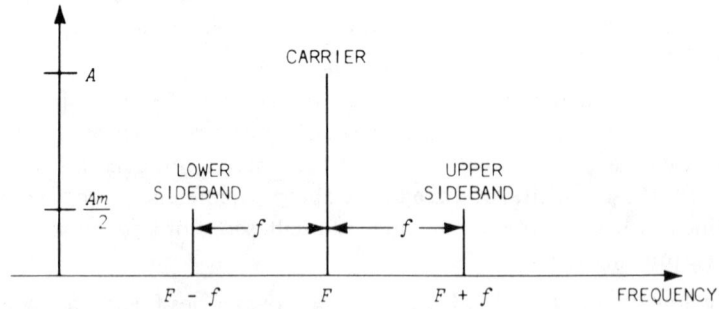

Figure 4-2 Frequency Domain Appearance of Amplitude Modulation

In the case of Figure 4-2, % modulation = 2 · (1/4) · 100 = 50%.

Besides the standard carrier with double sideband AM, systems that eliminate the carrier, or the carrier and one sideband, are also utilized. These are:

1. Reduced- or suppressed-carrier AM.

2. Single-sideband (SSB) AM.
3. Vestigial-sideband AM.

The rationale for use of these systems stems from a desire to reduce transmitter power requirements and to utilize more efficiently the available frequency space.

The suppressed-carrier signal, as the name implies, consists of two sidebands with a greatly reduced or attenuated carrier. Normally the carrier contains at least two-thirds of the transmitted power. The suppressed-carrier technique permits a reduction in transmitted power without reducing the size of the intelligence-bearing sidebands. The technique, while reducing transmitter power requirements, calls for a more complicated receiver design, since the carrier has to be reinserted to avoid distortion.

In single-sideband transmission, the usual practice is to eliminate one sideband and the carrier, although elimination of only the sideband is also called single sideband. Eliminating one sideband cuts the transmitted spectral width in half, thus conserving frequency space. As in suppressed carrier, SSB requires a more complex receiver since the missing sideband has to be reinserted by generating a mirror image of the transmitted sideband.

Sometimes, to ease network complexity, one sideband is merely reduced in amplitude rather than eliminated. This is particularly true when the information contains extremely low frequencies. Such an arrangement is called vestigial sideband. The best known example of vestigial-sideband transmission is in television. Here the vestigial sideband occupies about one-sixth the frequency space of the unattenuated sideband, thus conserving broadcast power and frequency space.

4.3. ANGLE MODULATION

As in amplitude modulation, it is not necessary that the signals be sinusoidal for angle modulation. However, for ease of analysis we shall confine the analysis to sinusoids.

As the name implies, in angle modulation it is the angle, rather than the amplitude of the sinusoidal carrier, $A \sin(2\pi Ft + \alpha)$, that is varied. There are essentially an infinite number of ways in which the angle of the carrier, $2\pi Ft + \alpha$, can be made to vary by the modulating wave, $B \cos 2\pi ft$. The two prevalent systems are phase modulation (PM) and frequency modulation (FM). In the former, the phase of the carrier is made to vary linearly with the modulating signal, while in the latter it is the frequency of the carrier that is made to vary in accordance with the modulating wave. For sinusoidal modulation, FM and PM are not much different, since instantaneous frequency is the derivative of phase and the derivative of a sinusoid is a sinusoid. Thus,

$$\text{instantaneous frequency} = \frac{1}{2\pi} \frac{d\theta}{dt} \qquad (4\text{-}8)$$

When dealing with a single sinusoid of the form A sin 2πFt, the angle θ = 2πFt, and

$$\frac{1}{2\pi}\frac{d\theta}{dt} = \frac{1}{2\pi}\frac{d}{dt} 2\pi Ft = F \qquad (4\text{-}9)$$

This shows that the definition of instantaneous frequency, Expression (4-8), is in agreement with the conventional understanding of the word.

Now consider PM and FM in more detail. For phase modulation, let the carrier be a sinusoid of the form a = A sin(2πFt + α), and let the modulating wave be represented by B cos 2πft. Phase modulation means that the phase of the carrier, which in unmodulated form is given by α, is modified by the modulating wave, resulting in a new phase of (α_0 + Δα cos 2πft). The complete expression is

$$a = A \sin\ 2\pi Ft + (\alpha_0 + \Delta\alpha \cos 2\pi ft) \qquad (4\text{-}10)$$

The quantity Δα cos 2πft is called the phase deviation and is expressed in radians. To reiterate — in PM the phase of the carrier is made to vary in accordance with the instantaneous amplitude of the modulating waveform resulting in a modulated waveform as given by Expression (4-10).

Now consider FM, which differs little from PM as will be shown. Since FM is more commonly used, it will be examined in much greater detail.

Let the carrier be of the form a = A (2πFt + α) and the modulating waveform B cos 2πft. Frequency modulation means that the instantaneous frequency of the carrier is modified in accordance with the instantaneous amplitude of the modulating waveform.

Combining the definition of instantaneous frequency from Equation (4-9) with the frequency of the carrier F and the form of the modulating waveform cos 2πft, the frequency of the modulated waveform is

$$\frac{1}{2\pi}\frac{d\theta}{dt} = F + \Delta F \cos 2\pi ft \qquad (4\text{-}11)$$

Equation (4-11) simply states that the instantaneous frequency of an FM signal is the sum of the carrier frequency and a term that has the form of the instantaneous amplitude of the modulating waveform, where θ stands for the phase of the modulated waveform that is given by a = A sin θ. To get the final equation for an FM signal, it is necessary to solve for θ. Integrating Equation (4-11) produces

$$\int d\theta = \int 2\pi F dt + \int 2\pi \Delta F \cos(2\pi ft) dt \ \ \text{or}$$

$$\theta = 2\pi Ft + \frac{\Delta F}{f} \sin 2\pi ft + \theta_0 \qquad (4\text{-}12)$$

Modulation Theory

The final result is that the FM wave is of the form

$$a = A \sin \theta = A \sin \left(2\pi Ft + \frac{\Delta F}{f} \sin 2\pi ft + \theta_0 \right) \quad (4\text{-}13)$$

The factor ΔF is called the peak frequency deviation, while $\Delta F \cos 2\pi ft$ is the frequency deviation. Frequency deviation means deviation with respect to the carrier frequency F.

It will be observed that Equations (4-13) for FM and (4-10) for PM are of the same form except for the factor $1/f$ in Equation (4-13). The major difference between FM and PM is that PM has increasing frequency deviation for increasing modulating frequency whereas FM has constant frequency deviation independent of the modulating frequency. The difference between FM and PM is particularly noticeable for multitone modulation where the ratios of the deviations at different frequencies is different for FM and PM.

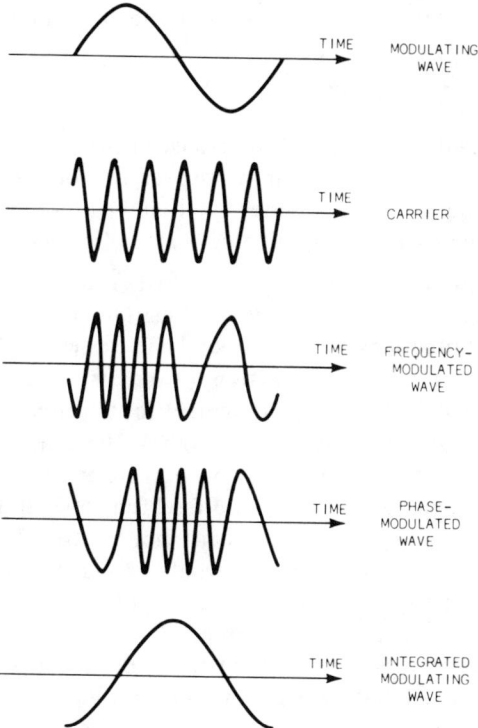

Figure 4-3 Time Domain Appearance of Angle Modulation

Figure 4-3 shows the time domain appearance of FM and PM. Note that the phase-modulated waveform has the appearance of frequency modulation of the integrated modulating waveform. This follows from the fact that instantaneous frequency is the differential of phase or, conversely, phase is the integral of instantaneous frequency. Since, for sinusoids, integration and differentiation merely involve a phase shift, it follows that, except for a change in deviation, FM and PM are the same for a sinusoidal modulating wave.

At this point, a discussion of Bessel functions is necessary for the frequency domain description of FM.

4.4. BESSEL FUNCTIONS

As discussed in Chapter 2, the circular trigonometric functions, $\sin \theta$ and $\cos \theta$ are the solution to the differential equation

$$\frac{d^2y}{dt^2} + \omega^2 y = 0 \qquad (4\text{-}14)$$

Similarly, Bessel functions are the solution of the differential equation

$$\frac{d^2y}{dt^2} + \frac{1}{t}\frac{dy}{dt} + \left(1 - \frac{p^2}{t^2}\right)y = 0 \qquad (4\text{-}15)$$

where p is a constant. While sinusoids, because of their long usage, appear simple and obvious, Bessel functions appear mysterious and forbidding. This need not be so. While deriving the various Bessel-function relationships* is not necessary, a discussion of the meaning of terminology can be tremendously helpful.

Just as the solution of the circular Equation (4-14) consists of two functions, a sine and a cosine, so the solution of Bessel Equation (4-15) consists of two functions called Bessel functions of the first kind and Bessel functions of the second kind. Of most interest are Bessel functions of the first kind, which are designated by the letter J. An equivalent statement for the circular trigonometric functions would be: of most interest are cosines. There are two parameters, ω and t, associated with the cosine. Likewise, there are two parameters, called the order and argument, associated with Bessel functions. In Bessel-function language, $\cos \omega t$ would be called a circular trigonometric function of the first kind of order ω and argument t. $\sin \omega t$ would be a circular trigonometric function of the second kind of order ω and argument t. While in circular trigonometric functions the order and argument appear as a product, in Bessel functions they are separated as follows: $J_p(t)$ means Bessel function of the first kind of order p and argument t. Just as for the circular functions, there is no restriction on how large the order or argument of Bessel functions can get. These are also not restricted to integral values, though for FM applications the interest is in integer multiples of the order p.

---

*See, for example, Whittaker & Watson, *Modern Analysis*.

Modulation Theory

Bessel functions are undulatory. But unlike the circular functions, the period, as measured between zero crossings, is not constant. Figure 4-4(A) is a graph of the first eight orders of Bessel functions of the first kind. By contrast, cos ωt, Figure 4-4(B), has constant period and constant peak amplitudes.

The value of a Bessel function is much more difficult to calculate than that for a circular function. The result is usually obtained from an infinite series such as

$$J_p(t) = \frac{t^p}{2^p p!} \left(1 - \frac{t^2}{2(2p+2)} + \frac{t^4}{2 \cdot 4(2p+2)(2p+4)} \cdots \text{ for positive integers of } p \right) \quad (4\text{-}16)$$

For the Bessel function of the first kind, order zero, Equation (4-16) becomes

$$J_0(t) = 1 - \frac{t^2}{4} + \frac{t^4}{2 \cdot 4 \cdot 8} \cdots \quad (4\text{-}17)$$

Fortunately, there are available many fine tables of Bessel functions, so that these can now be used almost as routinely as the circular trigonometric functions.

Bessel functions and the circular trigonometric functions are related. For example, at very large values of the argument t,

$$J_p(t) = \sqrt{\frac{2}{\pi t}} \cos\left(1 - \frac{p\pi}{2} - \frac{\pi}{4}\right) \quad (4\text{-}18)$$

That is, the larger the argument the closer does the Bessel function resemble a decaying circular function. Bessel functions and the circular trigonometric functions are even more fundamentally related to each other, since it can be shown that

$$\cos t = J_0(t) - 2\left[J_2(t) - J_4(t) + J_6(t) \ldots\right] \quad (4\text{-}19)$$

$$\sin t = 2\left[J_1(t) - J_3(t) + J_5(t) \ldots\right] \quad (4\text{-}20)$$

The fact that a sinusoid can be expanded as a series of Bessel functions should cause no surprise. Bessel functions are orthogonal; hence, as discussed in Chapter 2, other functions including sinusoids are expandable as a series of Bessel functions.

Of major importance in FM theory is that a sinusoid with a sinusoidal modulation angle is expandable as a series of sinusoids with Bessel function coefficients. The formulas that are related to Equations (4-19) and (4-20) are

$$\cos(t \sin \theta) = J_0(t) + 2\left[J_2(t) \cos 2\theta + J_4(t) \cos 4\theta \ldots\right] \quad (4\text{-}21)$$

$$\sin(t \sin \theta) = 2\left[J_1(t) \sin \theta + J_3(t) \sin 3\theta \ldots\right] \quad (4\text{-}22)$$

$$\cos(t\cos\theta) = J_0(t) - 2\left[J_2(t)\cos 2\theta - J_4(t)\cos 4\theta + J_6(t)\cos 6\theta \ldots\right] \quad (4\text{-}23)$$

$$\sin(t\cos\theta) = 2\left[J_1(t)\cos\theta - J_3(t)\cos 3\theta + J_5(t)\cos 5\theta \ldots\right] \quad (4\text{-}24)$$

Finally, an approximation useful in narrowband FM calculations is: For small arguments (t < 0.5), the zero order and first order Bessel functions of the first kind are related to the argument as follows:

$$J_0(t) \approx 1$$
$$J_1(t) \approx \frac{t}{2} \quad \text{for } t < 0.5 \quad (4\text{-}25)$$

4.5. THE FM SPECTRUM

The previously derived form of the FM wave is

$$a = A\sin\left(2\pi Ft + \frac{\Delta F}{f}\sin 2\pi ft + \theta_0\right) \quad (4\text{-}26)$$

where

F is the carrier frequency,

f is the modulation frequency,

A is the carrier amplitude,

ΔF is the peak deviation, and

$\Delta F/f$ is called the modulation index.

Using Equations (4-21) through (4-24), it can be shown that the FM wave Equation (4-26) is equivalent to an infinite series of sinusoids with Bessel coefficients as follows:

$$\begin{aligned}
a &= A\sin\left(2\pi Ft + \frac{\Delta F}{f}\sin 2\pi ft + \theta_0\right) \\
&= A\left\{J_0\left(\frac{\Delta F}{f}\right)\sin\left(2\pi Ft + \theta_0\right) + J_1\left(\frac{\Delta F}{f}\right)\sin\left[2\pi(F+f)t + \theta_0\right]\right. \\
&\quad - J_1\left(\frac{\Delta F}{f}\right)\sin\left[2\pi(F-f)t + \theta_0\right] + J_2\left(\frac{\Delta F}{f}\right)\sin\left[2\pi(F+2f)t + \theta_0\right] \\
&\quad + J_2\left(\frac{\Delta F}{f}\right)\sin\left[2\pi(F-2f)t + \theta_0\right] + J_3\left(\frac{\Delta F}{f}\right)\sin\left[2\pi(F+3f)t + \theta_0\right] \\
&\quad \left. - J_3\left(\frac{\Delta F}{f}\right)\sin\left[2\pi(F-3f)t + \theta_0\right] + \ldots\right\} \quad (4\text{-}27)
\end{aligned}$$

Equation (4-27) gives the frequency distribution, or spectrum, of a frequency-modulated wave. There are several important conclusions regarding the FM spectrum which can be drawn from Equation (4-27).

1. The FM spectrum consists of a set of discrete sinusoids.

Modulation Theory

2. These sinusoids appear at carrier frequency F and sidebands on either side of the carrier spaced the modulating frequency f apart.
3. There is no end to the sinusoids; theoretically, the FM spectrum has infinite frequency distribution.
4. The amplitudes of the carrier component and the various sidebands are determined by the product of the original carrier amplitude A and the value of a Bessel function. The order of the Bessel function corresponds to the sideband number counting the carrier as number zero. The argument of the Bessel functions is the modulation index $\Delta F/f$.
5. Since the amplitude of the carrier component is modified by the factor $J_0(\Delta F/f)$, it follows that the carrier component of the modulated wave is smaller in amplitude than the unmodulated carrier. As a matter of fact, the carrier component can actually go to zero. This is called a carrier null and happens when $J_0(\Delta F/f) = 0$. The first carrier null occurs at a modulation index of 2.4, as can be seen by the zero crossing of the $J_0(t)$ curve in Figure

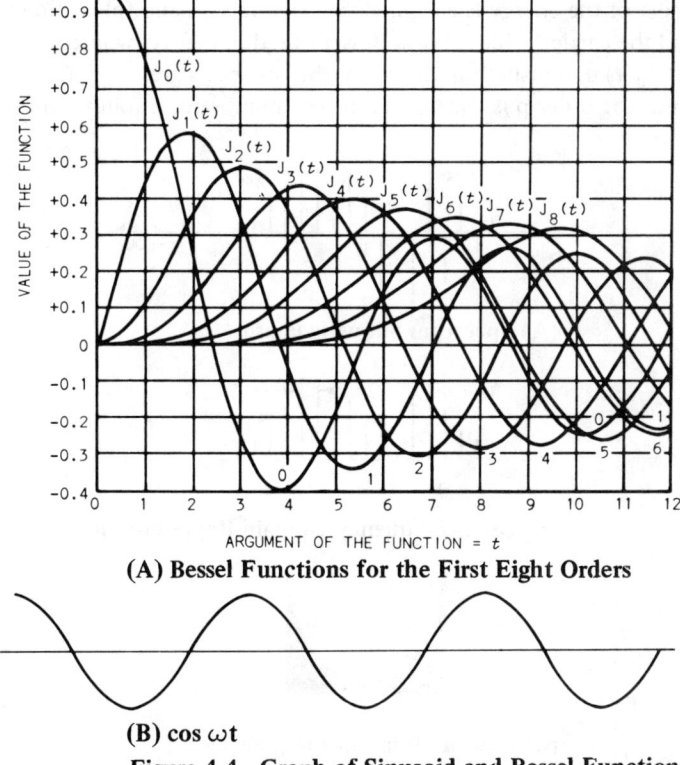

(A) Bessel Functions for the First Eight Orders

(B) cos ωt

Figure 4-4 Graph of Sinusoid and Bessel Functions

4-4(A). These Bessel zeros are used in determining the frequency deviation as discussed in the section on measurements. FM is a constant-energy process where energy is removed from the carrier and supplied to the sidebands. Thus, the energy of an FM wave is constant regardless of the degree of modulation. This is in contrast to AM, where the carrier amplitude is constant and the modulation process adds energy to the wave.

Figure 4-5 shows a typical FM spectrum. There are two important points that should be indicated. One is that while a spectrum analyzer, being insensitive to phase, will show the FM spectrum as in Figure 4-5(B), the actual spectrum is as shown in Figure 4-5(A). Here is shown the fact that the odd upper and lower sidebands are 180° out of phase with respect to each other. This is demonstrated in Equation (4-27) by the alternating positive and negative signs associated with the odd numbered sidebands. This out-of-phase characteristic will be used in deriving the spectrum of combined AM and FM and later in the section on applications as a means for differentiating between AM and narrowband FM. The second point is that, while theoretically the FM spectrum does go on ad infinitum, most of the energy is confined to a frequency band (plus and minus ΔF) around the carrier. This follows Bessel function theory, where it can be shown that $J_p(t)$ diminishes rapidly when the order p is greater than the argument t. Since the order p is equal to harmonic number n, it follows that as the

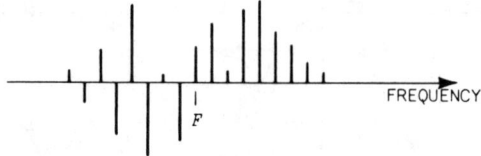

(A) Spectrum Showing Phase

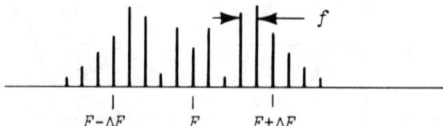

(B) Usual Frequency Domain Representation

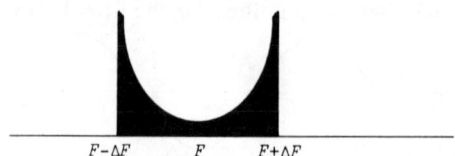

(C) Low Modulation Frequencies

Figure 4-5   FM Spectrum

Modulation Theory

modulating frequency f gets smaller it takes more harmonics to cover the frequency range ΔF. Hence, as $f \to 0$ the frequency spectrum acquires sharp demarcation lines at plus and minus ΔF around F as shown in Figure 4-5(C).

4.6. COMBINED MODULATION

In multitone AM, each modulating frequency can be treated individually as if the others were not there. Hence, the spectrum of multitone AM is just the sum of the individual single-tone spectra. In multitone FM, there is an interaction between the several modulating signal frequencies, creating additional sidebands than is apparent by treating each tone individually. The mathematics for multitone FM can get quite complicated and will not be reproduced here. A major difference between the spectra of single-tone and multitone FM is that while in the former the sideband distribution is symmetrical about the carrier, in the latter it need not be. While an absolute rule is difficult to formulate, because of the complexity of the situation, it has generally been found that symmetrical modulating waveshapes create symmetrical spectra while unsymmetrical modulating waveshapes create unsymmetrical spectra. Thus, unless the modulating waveform is a pure sinusoid, it is possible to get an unsymmetrical spectrum in FM. In multitone, as in single-tone FM, the total energy is constant regardless of the degree of modulation. Hence, in multitone FM, as the number of sidebands is increased, the carrier component is decreased.

Simultaneous AM and FM is usually an accidental, or incidental, phenomenon rather than a deliberate form of modulation. This form of modulation usually occurs when it is desired to obtain AM. Somehow the carrier oscillator frequency is pulled by the modulating signal, introducing a small amount of identical FM along with the AM. The result is AM along with narrowband (low-modulation-index) FM at the same modulating frequency as the AM. Let us, therefore, consider the theoretical spectrum of AM combined with narrowband FM.

The AM spectrum consists of a carrier and two sidebands, as given by Equation (4-5). The FM spectrum consists of a carrier and an infinity of sidebands, as given by Equation (4-27). However, as previously indicated, the amplitude of the FM sidebands falls off very rapidly outside of the peak deviation interval $\pm \Delta F$. In narrowband FM, where ΔF is considerably less than the modulating frequency f, higher order sidebands fall off so rapidly that all but the first sidebands can be ignored.

Thus, the spectrum is given by

$$a = A \left\{ J_0 \left(\frac{\Delta F}{f} \right) \sin(2\pi Ft + \theta_0) + J_1 \left(\frac{\Delta F}{f} \right) \sin \left[2\pi(F + f)t + \theta_0 \right] \right.$$
$$\left. - J_1 \left(\frac{\Delta F}{f} \right) \sin \left[2\pi(F - f)t + \theta_0 \right] \right\} \qquad (4\text{-}28)$$

which is simply Equation (4-27) but ignoring all but the first sideband. Narrowband FM differs from AM in that the initial phase of one sideband is 180° different than the initial phase of the other. These are difficult to distinguish on a spectrum analyzer because the spectrum analyzer is insensitive to phase. A technique for distinguishing between narrowband FM and AM is discussed in the section on applications in this chapter. Combined AM and narrowband FM, however, has a distinctive spectrum that is easily identified. Consider, for example, the case illustrated in Figure 4-6. Here are shown the spectra of AM, narrowband FM and a combination of the two. While it is true that the spectrum ana-

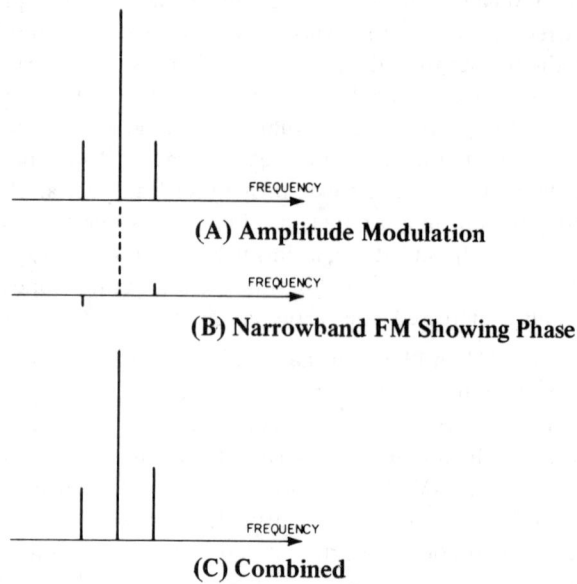

Figure 4-6 Combining AM and Narrowband FM

lyzer is insensitive to phase, the displayed combined spectrum has to be considered in accordance with the principle of superposition, where phase has to be accounted for. The result, shown in Figure 4-6(C), is a distinctive spectrum where one sideband is larger than the other.

One point that needs clarification is that, while the sidebands are added, the carrier is shown as constant amplitude. This is because the AM and FM spectra actually share the same carrier. In the FM spectrum, the carrier is shown dotted, because in narrowband FM there is so little energy in the sidebands that the carrier may be assumed to be unaffected, and of course in AM, the carrier is

Modulation Theory

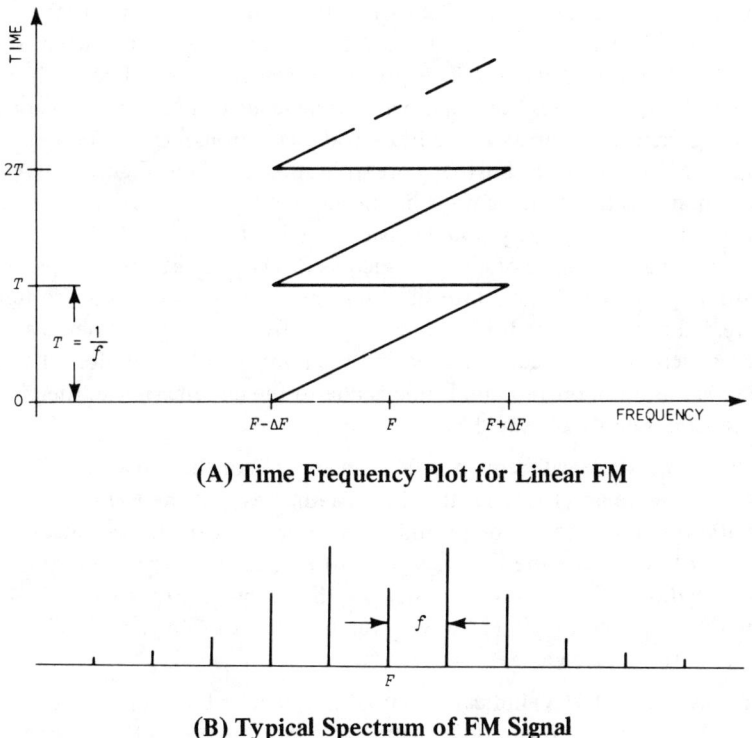

(A) Time Frequency Plot for Linear FM

(B) Typical Spectrum of FM Signal

Figure 4-7 Dual Representations of FM Signal

unaffected regardless of the percentage modulation. It is conceivable that the equipment is so constructed that the carrier level is affected by the incidental FM. This can only be determined from a knowledge of the actual equipment. In the absence of such knowledge, it is assumed that the carrier remains unaffected, which is the prevalent case.

4.7. TIME DOMAIN AND FREQUENCY DOMAIN

Many people have difficulty in accepting the validity of spectrum-analyzer-derived results because many of these results seem contrary to common sense. While this problem was already considered in a general way in Chapter 2, it is worthwhile to resolve the specific case of FM. FM is of special interest because nowhere is the paradox between the spectral distribution and common sense more obvious than in this case.

Consider linear frequency modulation such as obtained from a sinusoidal carrier modulated with a sawtooth signal. The time-frequency relationship for the FM wave is shown in Figure 4-7(A). The carrier at a frequency F is made to vary its frequency in a linear fashion from $F - \Delta F$ to $F + \Delta F$, where ΔF is what is normally called the peak deviation. This process is repeated every T seconds, or the FM rate is $f = 1/T$. It does not require any knowledge of FM theory to conclude that during the time interval T, the FM signal goes through every frequency between $F - \Delta F$ and $F + \Delta F$. Furthermore, it is apparent from Figure 4-7(A) that the instantaneous frequency always lies in the $F - \Delta F$ to $F + \Delta F$ interval. This rationale is behind the highly popular sweep-testing technique. In sweep testing, the transfer characteristics of circuits, such as filters, are obtained by feeding a constant-amplitude FM signal into the circuit undergoing the test and observing the detected output. Since it is assumed that all frequency components are equally present at the circuit input within the limits of FM deviation, it follows that any variation in output amplitude is due to the circuit. Hence, the circuit characteristic is easily obtained.

Consider now a standard frequency-domain analysis of the FM signal. This result is shown in Figure 4-7(B). Here, the theory indicates that the FM wave consists of specific frequency components and nothing in between. These sinusoids, that appear to compose the FM wave, consist of a carrier component and sidebands spaced on either side of the carrier with a frequency spacing of $f = 1/T$. The two descriptions of the FM wave given in Figures 4-7(A) and 4-7(B) appear to be in conflict in two areas —

1. Whereas Figure 4-7(A) indicates that all frequencies between $F - \Delta F$ and $F + \Delta F$ are present, Figure 4-7(B) indicates that there is only energy at very specific frequencies and nothing anywhere else.
2. While it is clear from Figure 4-7(A) that no instantaneous frequency exists outside the frequency range $F - \Delta F$ to $F + \Delta F$, it is equally obvious from standard FM theory that there is no end to the sidebands.

So we have the paradox that the spectrum analyzer seems to indicate that there is nothing where logic says there should be something and, conversely, the spectrum analyzer indicates energy at frequencies where logically there should be nothing. Actually, both interpretations are correct because they apply to different circumstances.

First, it should be recognized that for a meaningful, practical discussion it is necessary to consider the behavior of circuits because, in the final analysis, there is only one way to determine whether there is or isn't energy at a specific frequency — this is by means of a measurement using real equipment made up of circuits. What is the difference between the circuits that would respond in the manner implied in Figures 4-7(A) and 4-7(B), and why do these different circuits give different frequency domain results for FM?

As discussed elsewhere, there is only one way in which a spectrum analyzer is

Modulation Theory

made to resolve or display individually the separate frequency components of a signal; this is by making the resolution bandwidth narrower than the frequency separation between signals. Hence, the difference is that Figure 4-7(A) implies a relatively wideband circuit while Figure 4-7(B) implies a relatively narrowband circuit. These circuits give different results because in one case the transient response is negligible while in the other case much of the output is due solely to the transient.

As discussed in Chapter 2, Response of Circuits to Signals, a transient response need not be of short time duration. When dealing with high-Q narrowband circuits, the transient can be quite long. Hence, it is logical that the transient response of a narrowband filter should contribute more to the total output than is the case for a wideband filter. Now, consider the fact that the stimulus to the filter is repeated once every T seconds. This means that the transient response, whatever its characteristics, is repeated at intervals of T seconds. If the filter has a time constant such that the transient response does not die down too much in T seconds, it follows that the transient output never disappears since it is regenerated in a shorter time interval than it takes to die out. Time constants of filters are basically proportional to the inverse of the bandwidth. Hence, if the filter bandwidth is narrow enough to separate the several frequency components, meaning that the bandwidth is less than the FM rate f, it follows that the time constant is on the order of T and the transient is reconstituted faster than it dies out.

It is not necessary to go through the derivation of the network response leading to the remarkable fact that the total output of a narrowband filter with an FM input has energy only at discrete frequencies. While the input consists of a time-variable signal going through all the frequencies between $F - \Delta F$ and $F + \Delta F$ the output of a narrowband filter, consisting of the combined transient and steady-state response and averaged over one FM cycle of interval T, contains no energy except at the frequencies indicated by accepted FM theory. At all frequencies, except the very special ones, currents flow in the filter in such a way that on the average there is no energy transfer. Furthermore, if the narrowband filter is outside the frequency range of the input, it is still possible to get an output. This is because a filter will respond with a transient to an input outside its frequency range. Normally, the transient dies down very quickly and can be ignored. However, if the stimulus is repeated at a fast enough rate and in appropriate synchronism with the filter frequency, one gets what looks like a continuous input. This is analogous to a swing pushed at a rate in synchronism with its natural frequency, resulting in continuous large oscillations.

When the filter bandwidth is large, as compared to the FM rate, the transient response is negligible so that the output has the same frequency characteristics as the input.

As was amply discussed in Chapter 2, it is not the purpose of this book to resolve the question of whether the spectral components are a part of the signal or are generated by the circuit. The important thing to remember is that real, physically realizable, linear, time-invariant circuits behave as if spectral components exist, and this is what the spectrum analyzer will show.

4.8. EXAMPLES

1. An amplitude modulated wave has a spectrum consisting of a carrier and two sidebands which are one one-hundredth the size of the carrier — What is the percentage modulation? From Equation (4-7),

$$\% \text{ modulation} = \frac{2 \cdot A_{sideband}}{A_{carrier}} \, 100 = \frac{2 \cdot \frac{1}{100}}{1} \, 100 = 2\%$$

2. Would this be a routine measurement on an oscilloscope? From Equation (4-6), $m = (K - 1)/(K + 1) = 2/100$, resulting in $K = 1.04$. An amplitude ratio of 1.04 is very difficult to measure on an oscilloscope.

3. Given a wave which is frequency modulated at a 10 kHz rate. The spectrum shows the first carrier null. What is the deviation? The first carrier null occurs at a modulation index of 2.4; $\Delta F/f = 2.4$, $\Delta F = 2.4 \cdot 10 = 24$ kHz.

4. An FM spectrum shows a 10 kHz sideband spacing and the following relative amplitudes of its components; Carrier, one; first sideband, zero; second sideband, one; third sideband, one; fourth sideband, six tenths. What is the deviation? From Figure 4-4(A), $J_1(t)$ has zeros at an argument of about 3.8, 7, 10.2 . . . . At only one of these does $J_0(t)$, $J_2(t)$ and $J_3(t)$ have the same magnitude — at a modulation index of 3.8. Here the various magnitudes are:

$J_0(3.8) \cong -0.4$

$J_1(3.8) \cong 0$

$J_2(3.8) \cong 0.4$

$J_3(3.8) \cong 0.4$

$J_4(3.8) \cong .25$

These are in the ratios of 1/1, 0/1, 1/1, 1/1, .625/1. Hence the modulation index is 3.8 and the deviation is $3.8 \cdot 10 = 38$ kHz.

5. Given narrowband FM at a 10 kHz rate, the sidebands are one-fiftieth the amplitude of the carrier. What is the deviation? From Equation (4-25), $J_0(t) \cong 1$, $J_1(t) \cong t/2 = 1/50$, and $t = 1/25$. The deviation is $(1/25) \, 10$ kHz = 400 Hz.

Modulation Theory 87

4.9. EXERCISES

4-1. A carrier is pulsed on and off to form a squarewave having the same peak-to-peak amplitude as the original carrier. Using the unmodulated carrier as the reference, what is the amplitude of the carrier component and first three sidebands for the pulsed RF signal. (Hint, use AM theory, Fourier series for squarewave, and frequency shift Theorem No. 8 in Table 3-1.)

4-2. Derive Equation (4-27) from the basic frequency modulation Equation (4-26) and the sinusoid Bessel function Equations (4-21) to (4-24).

4-3. Given an amplitude-modulated waveform 70% modulated by a single sinewave,
   a. What is the voltage amplitude of the sidebands compared to the carrier?
   b. How large is E_{max} and E_{min} as compared to the carrier?

4-4. Compute the modulation index corresponding to the fifth carrier null in FM and compare to the value given in Table A-2. (Hint, use the large argument approximation of $J_0(t)$.)

4-5. A carrier is frequency modulated by a 1 kHz sinewave 2 volts in amplitude resulting in a 4 kHz deviation. If the unmodulated carrier has an amplitude of 10 volts,
   a. What are the amplitudes of the carrier and first three sidebands?
   b. What will these amplitudes be if the modulating voltage is reduced to 1 volt? (Assume a linear modulator.)

4-6. A spectrum consists of a carrier of amplitude 1 volt and 2 sidebands spaced 1 kHz from the carrier. The lower sideband amplitude is 0.09 volts and the upper sideband is 0.11 volts. Determine the parameters associated with this modulation (e.g., % if AM, deviation if FM, etc.).

Chapter 5
The Sweeping-Signal Spectrum Analyzer

5.1. INTRODUCTION

As discussed in Chapter 1, the superheterodyne, or sweeping-signal spectrum analyzer, operates on the principle of the relative movement in frequency between the signal and a filter. The important parameter is the relative frequency movement. It does not matter whether the signal is stationary and the filter changes frequency nor whether the filter is stationary and the signal is made to change frequency.

Figure 5-1 shows the spectral representation obtained in such a system. Figure 5-1(A) represents a spectrum composed of three discrete-frequency CW signals and a continuous dense spectrum in the middle. This spectrum is passed through a filter having the gain characteristic shown in Figure 5-1(B). The filter and spectrum have a relative frequency translation as indicated by the arrows of opposite sense. The resultant display, shown in Figure 5-1(C), has the units of frequency, ω, but takes a real time, t, to occur. Some of the fine detail of the theoretical spectrum, shown in Figure 5-1(A) as a Fourier transform $F(\omega)$, is lost in Figure 5-1(C) because of the finite frequency width and, hence, resolution of the filter. As the resolution filter gets narrower, the ideal and actual spectral representations get more alike until, when the filter has zero bandwidth (in effect, an impulse function), the ideal and actual representations become the same. The transformation of the ideal spectrum into the actual spectral representation by the relative frequency translation between filter and signal is known as convolution. Convolution was previously discussed in Chapter 3, and a further discussion will be found in the appendix to this chapter.

While the basic operation of the system is apparent from Figure 5-1, there are many ramifications, particularly with regard to the speed of relative-frequency translation, which are not at all obvious. Let us now consider some of the details of the sweeping-signal spectrum-analyzer system.

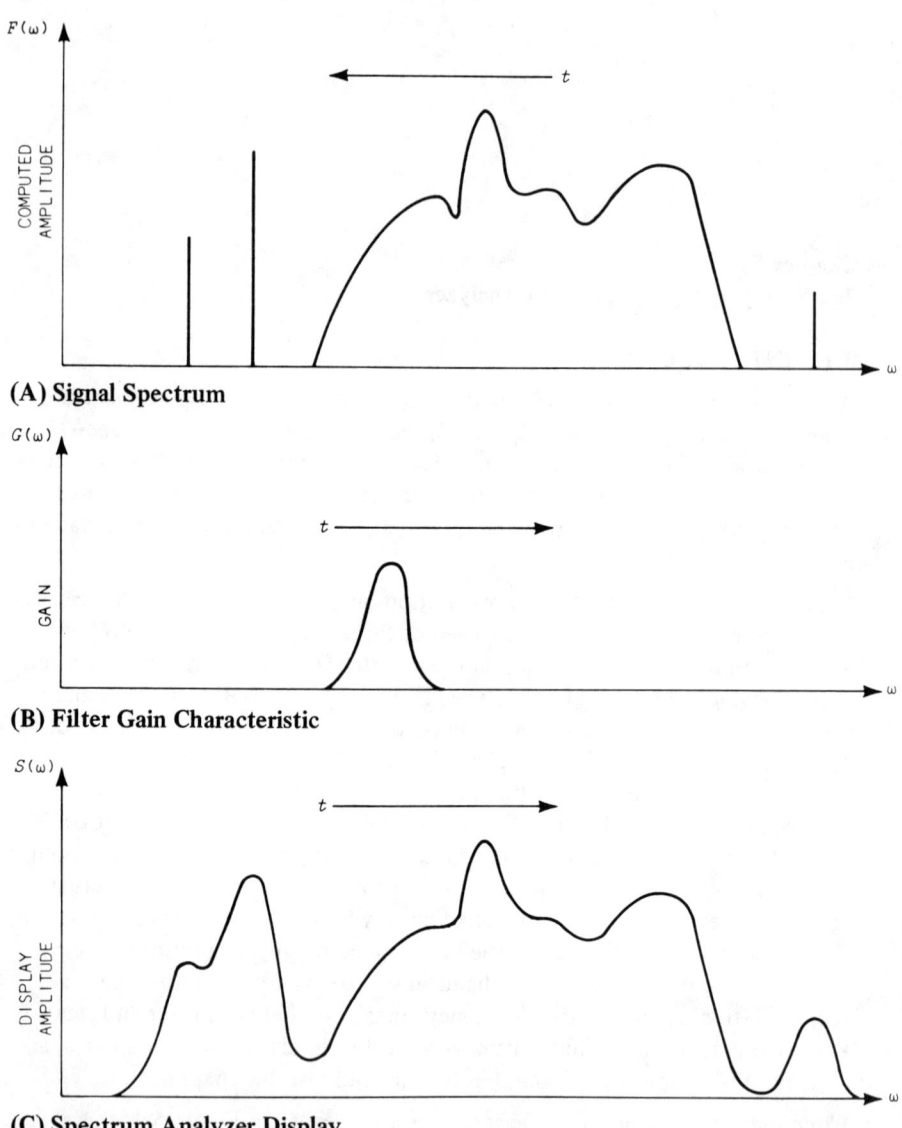

(A) Signal Spectrum

(B) Filter Gain Characteristic

(C) Spectrum Analyzer Display

Figure 5-1 Sweeping-Signal Spectrum Analyzer Spectrum Representation

5.2. THE CW RESPONSE

While there are many possible configurations (e.g., swept IF or swept front end) as discussed in another chapter, all spectrum analyzers of the type under discussion contain a mixer, sweeping oscillator, and resolution filter. The simplest

The Sweeping-Signal Spectrum Analyzer

arrangement, which is sufficient for the purpose of theoretical discussion, is shown in Figure 5-2. The time/frequency diagram for this system is shown in Figure 5-3. Here it was assumed that system operation is based on a mixer output composed of the difference frequency between local oscillator and signal. Likewise, it was assumed that the signal is composed of two discrete frequency components. The signal components at frequencies f_1 and f_2 are shown as straight lines having infinitesimal frequency width and infinite time duration.

A constant-frequency signal is converted to a frequency sawtooth by combining it in a mixer with a frequency sawtooth from the swept local oscillator. In the example, it was assumed that the mixer output consists of the difference frequency between the local-oscillator frequency sawtooth and the input. Other combinations, such as the sum of the frequencies, lead to similar diagrams. The display consists of pulses whose time position is determined by the time of intersection of the filter passband and the sweeping signal, and whose width is equal to the time interval during which the sweeping-signal frequency is within the filter passband. The bursts or pulses generated by the relative translation of signal and filter are pseudo-impulses representing the frequency-domain characteristics of the signal. While the time position of these pulses represents the input signal frequency and is determined by the incoming signal, the width τ of these pulses is determined solely by the spectrum-analyzer parameters. The width τ is equal to the time that the sweeping-signal frequency is within the passband of the filter, and from simple geometrical considerations is:

$$\tau = \frac{B}{S} T \qquad (5\text{-}1)$$

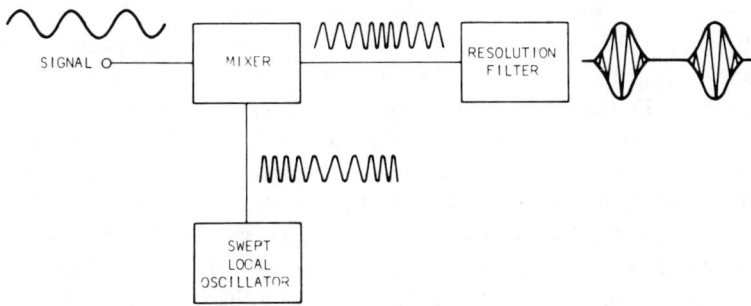

Figure 5-2   Basic Spectrum Analyzer Block Diagram

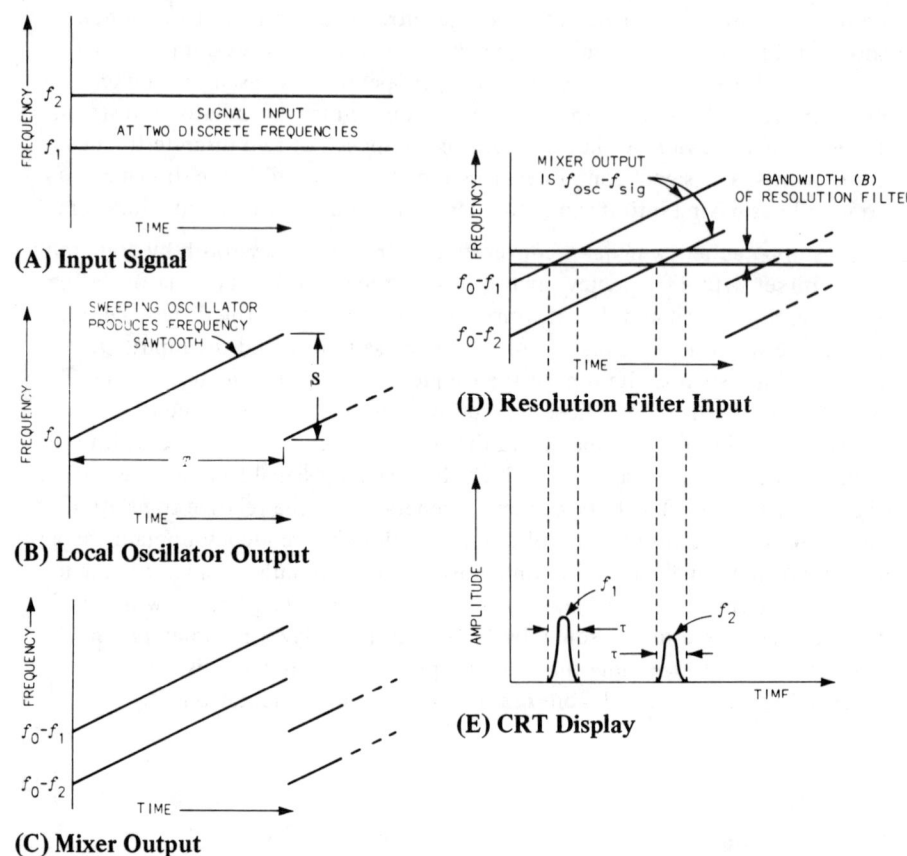

Figure 5-3 Time/Frequency Diagram, Sweeping-Signal Spectrum Analyzer

The sweep time T is the time it takes the electron beam to traverse the horizontal width of the CRT. Hence the physical width of τ in inches or centimeters does not change with changing T. The actual time duration of τ, however, given by Equation (5-1), is directly determined by the sweep time T. At low sweep time T, or with narrow resolution-filter bandwidth B, or with large span S, the pulse width τ can become quite small. For example, a full-screen sweep time of 1 ms (100 µs/DIV), a resolution bandwidth of 10 kHz, and a span of 10 MHz result in a burst at the filter output that is only 1 µs wide. Such a narrow pulse cannot be passed by a 10-kHz-wide filter without distortion. As with any pulse that is passed through a filter of insufficient bandwidth, the output is of lesser amplitude and greater time duration than the input, as illustrated in Figure 5-4. Since the distorted response is what appears on the CRT screen, the loss in amplitude shows a loss in sensitivity, and the apparent widening of the resolution

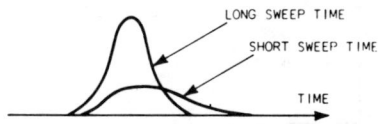

Figure 5-4 Resolution Distortion for Short Sweep Time

bandwidth shows a loss in resolution. Analytical expressions, relating the amount of loss to the sweep time T, to actual resolution bandwidth B and to span S, have been developed by many people. Some of the results are based on convolution techniques using the $e^{-1/4}$ bandwidth as the standard,* while others calculate the transient response using a 3-dB standard bandwidth along the lines discussed in Section 4.6, Volume XI of the Radiation Laboratory Series.† When the differences in bandwidth are accounted for, the final results are essentially the same for both methods. The ratio of apparent resolution bandwidth R to actual bandwidth B is

$$\frac{R}{B} = \left[1 + 0.195 \left(\frac{S}{TB^2}\right)^2\right]^{1/2} \tag{5-2}$$

where B is the 3 dB bandwidth.

The amplitude loss factor, α, is

$$\alpha = \left[1 + 0.195 \left(\frac{S}{TB^2}\right)^2\right]^{-1/4} \tag{5-3}$$

where B is the 3 dB bandwidth.

Figure 5-5 is a plot of Equations (5-2) and (5-3). While these equations are theoretically correct, they do not necessarily correspond to the behavior of actual equipment. The discrepancy between theory and practice arises because the theoretical results are based on a filter having a gaussian amplitude response and linear phase response which in practice is not necessarily the case. Differences between theoretical and actual phase response can be particularly important. As a matter of fact, with appropriate phase response pulse compression, such as is utilized in chirp radar, can be achieved so that resolution need not be degraded at high sweep rates.

---

*Batten, et al., "The Response of a Panoramic Receiver to CW and Pulsed Signals," *Proc. IRE*, June, 1954.

Chang, "On the Filter Problem of the Power Spectrum Analyzer," *Proc. IRE*, August, 1954.

†*Spectrum Analyzer Techniques Handbook*, Polarad Electronics Corp.

"Spectrum Analysis," *Application Note 63*, Hewlett Packard.

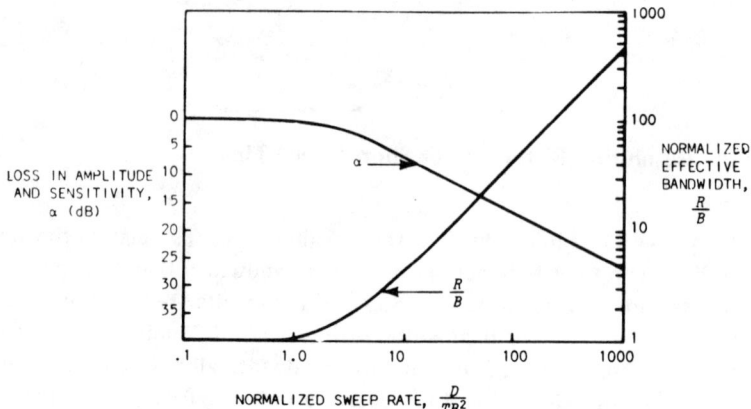

Figure 5-5 Loss in Sensitivity and Resolution as a Function of Sweep Rate

While pulse-compression spectrum analyzers are presently economically unfeasible, the theory is nevertheless sound. While Equations (5-2) and (5-3) are useful as a general guide to spectrum analyzer performance, there is no substitute for experimental data on the actual equipment in question.

Modern spectrum analyzers maintain an appropriate S, T, and B relationship automatically so as not to lose calibration. Formerly, it used to be the practice to increase the sweep time T until the display is no longer distorted. There are, however, situations when this is not possible. Such a situation might arise when the signal under investigation has a limited time duration so the complete analysis must be performed in less than a specified time interval. Under such circumstances, sweep time T and span S are usually fixed by the signal parameters and the only variable is the resolution bandwidth B.

Which value of actual bandwidth B results in minimum displayed resolution R? This is easily obtained by differentiating Equation (5-2) and letting dR/dB equal zero. Thus,

$$\frac{dR}{dB} = B - \left[\frac{0.195}{B^3}\left(\frac{S}{T}\right)^2\right] = 0 \tag{5-4}$$

From which it follows that

$$B_0 = \sqrt{\frac{1}{2.27}\frac{S}{T}} \tag{5-5}$$

where B_0 is the optimum bandwidth which, at a given setting of T and S, results in minimum (or optimum) displayed resolution bandwidth R_0.

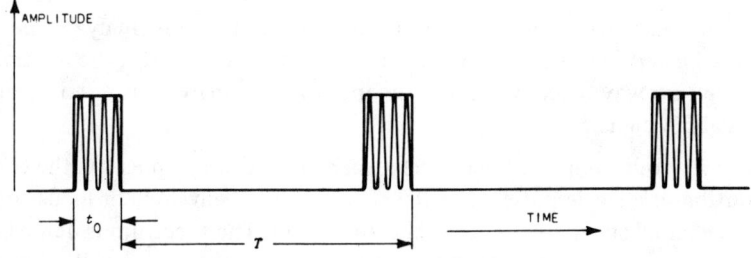

(A) Pulse Train in Time Domain

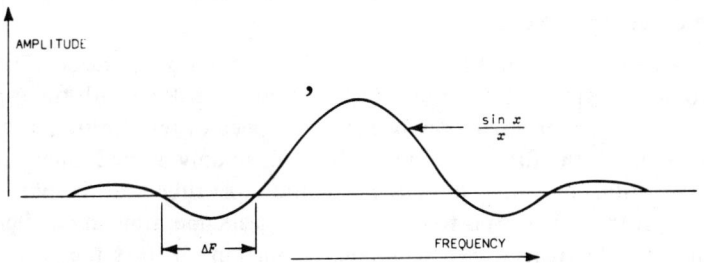

(B) Pulse Train in Frequency Domain

Figure 5-6 Pulsed Signal Analysis

When B_0 is substituted back into Equation (5-2), the result is that $R_0 = \sqrt{2}\; B_0$. Theoretically, then,

$$R_0 \cong \sqrt{\frac{S}{T}} \qquad (5\text{-}6)$$

any other value of B results in a larger R. R_0 and B_0 are generally known as optimum resolution and optimum resolution bandwidth respectively. While the optimum resolution is generally proportional to the square root of the ratio of span to sweep time, the proportionality constant need not necessarily be unity as given by Equation (5-6). Equation (5-6) is based on a Gaussian, linear phase response filter, which is not usually the case in practice.

5.3. PULSED SIGNALS

In Chapter 3, it was shown that the theoretical spectrum of a train of pulses has a (sin x)/x envelope as shown in Figure 5-6. This (sin x)/x curve can be interpreted in two ways. One interpretation is that the (sin x)/x is the envelope formed by the locus of the end points of the fundamental and harmonic sinusoids, which combine to produce the pulse train. The display consists of a set of vertical lines, each representing a sinusoid. If the spectrum analyzer span were so adjusted as to show only one of these lines on the CRT, the observer would have no way of knowing that the input to the analyzer is not a single sinusoidal CW signal.

Another way of looking at the (sin x)/x shape is that this represents the energy distribution of a single pulse. Here the curve is not an envelope or locus curve but is the actual shape of the spectral distribution. The spectrum is dense and continuous, and one no longer speaks of individual harmonics or CW responses. The detailed reasoning involved in these two interpretations will be found in Chapter 3. What is of interest here is the processes in the analyzer that lead to one or the other type of display.

Consider the dense spectrum first. Here, the object is to obtain the spectral distribution of a single pulse. Unfortunately, this cannot be done with the type of spectrum analyzer under discussion, because the speed of relative frequency translation between the filter and signal is limited, so only a small range of the frequencies of interest can be checked during the short time that the pulse exists. One way out of this dilemma is to use many analyzers operating in parallel and look simultaneously at the different frequency portions of the same pulse. Another way is to have many identical pulses at which a single analyzer can look sequentially. Here, many systems working a short time with one pulse have been traded for one system working a long time with many pulses. As long as the many pulses are identical and each succeeding measurement is independent of all previous measurements, the results of the two configurations will be identical. In considering dense spectra, the many-pulses, single-analyzer system is used. While the requirement that all pulses in the pulse train have the same characteristics can only be controlled by the source of the signal, the need for each successive measurement to be independent of all previous measurements is under the control of the spectrum analyzer user. All that is needed to make each measurement independent of all others is that all traces (e.g., electron beam deflections) of the previous measurements be dissipated in the interval between measurements. Since measurements are performed one per pulse, the measurement interval is the interpulse interval. Also, the memory of a circuit is essentially proportional to its time constant, so the shorter the time constant the less trace is left from previous measurements in a given time interval. Finally, time constants are inversely proportional to bandwidths and interpulse intervals are inversely proportional to repetition rates. Hence, for each measurement to be independent of all others, it is necessary that the resolution bandwidth be greater than the pulse-train repetition rate. Let us now consider what actually happens in circuits.

Figure 5-7 shows a time/frequency diagram similar to Figure 5-3, only the input consists of a pulse train. Each of the pulses, theoretically, has a broad spectrum

The Sweeping-Signal Spectrum Analyzer

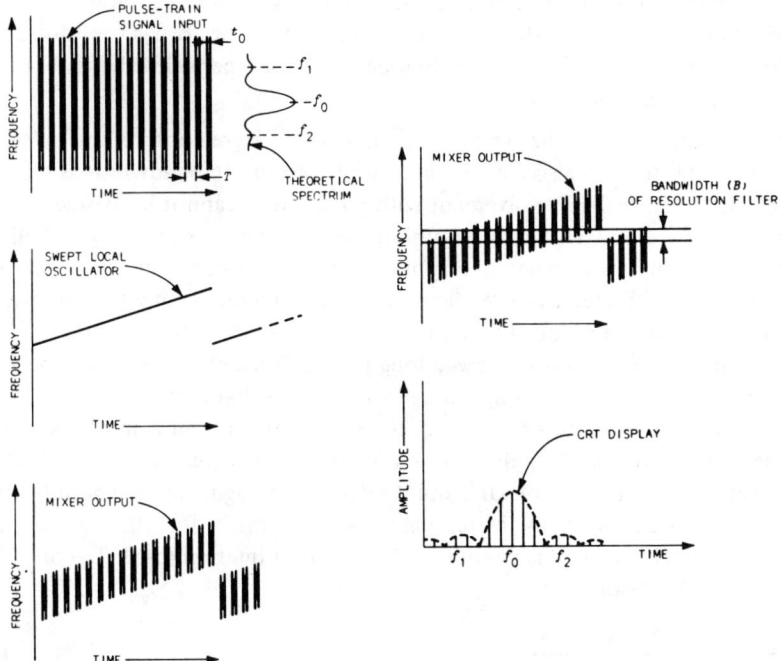

Figure 5-7 Time Frequency Domain for Pulsed Signal Analysis

so the signal exists at many frequencies simultaneously. However, the signal is not continuous in time, but occurs in narrow bursts, t_0 seconds in duration every T seconds. The distributed spectrum of each pulse is in turn moved in frequency by mixing with the sweeping local oscillator, resulting in the mixer output as shown in Figure 5-7. Every intersection between the mixed sweeping spectrum and the stationary filter results in an output. As long as the pulse rate $(1/T)$ is less than filter bandwidth B, each input pulse produces an output that is independent of all the other pulses so the resultant display has the shape of the theoretical single-pulse dense spectrum.

Each time a pulse occurs, there is an output. Hence, the composite output consists of lines, one line per pulse. This means that the total number of lines is equal to the number of pulses per sweep. As the sweep time is changed, the number of lines will change. These lines are not Fourier or spectral lines but are strictly determined by the number of pulses per sweep and are usually termed rep-rate lines.

Since the number of lines forming the spectrum depends on the sweep time, it is necessary that the sweep time be sufficiently long to generate enough pulse intercepts to define adequately the shape of the spectrum. It has been found experimentally that 10 lines per main lobe and 5 lines per side lobe is about the minimum that one can use.

Besides the requirement that resolution bandwidth be greater than the pulse repetition rate, there is also a constraint on the resolution bandwidth as a function of pulse width. The involvement with pulse width cannot be avoided, since, in truth there are no CW signals, only short pulsed sinusoids and long pulsed sinusoids. The matter of when a long pulse (seconds, minutes, hours or days in duration) can be treated as a CW signal can be considered from a time domain or frequency domain point of view, and both analyses lead to the same result. From a time domain point of view, a long pulsed sinusoid might as well be a continuous wave if it exists long enough to trace the shape of the resolution filter. From Figure 5-3 this means that the spectrum analyzer cannot distinguish between a CW signal and a pulse whose time duration is greater than $\tau = (B/S)T$, as is given by Equation (5-1). Offhand it might be thought that τ would be made as small as desired simply by decreasing the sweep time T. But there is an optimum beyond which there is trouble, as is given by Equation (5-5). $T = S/2.27B^2$. Substituting for T, one gets:

$$\tau = \frac{B}{S} \cdot \frac{S}{2.27B^2} = \frac{1}{2.27B} \tag{5-7}$$

This means that, from a time domain point of view, the demarcation line, between a pulse looking like a pulse or like a CW signal, is a pulse width about one-half of the inverse of resolution bandwidth. From a frequency domain point of view, a long pulse looks like a CW signal when the resolution bandwidth is sufficiently wide to encompass most of the spectrum energy.

Most of the pulse energy is in the main lobe, which has a frequency width of $2/t_0$, where t_0 is pulse width. Hence, we come to the conclusion that $t_0 B \approx 2$ is the demarcation line between a pulse-like spectrum and a CW-like spectrum. Naturally, a resolution bandwidth that is just on the border line will not permit the display of the fine detail of a pulsed spectrum. It has been found experimentally that for adequate detail the pulse-width bandwidth product should be less than one-tenth, thus

$$t_0 B \leqslant 0.1 \tag{5-8}$$

A major difference between a pulse-type response and a CW-type response is in the width of the pulse that the final amplifier has to pass. In the CW case, there is the relatively wide pulse, τ, due to the steady-state response of the resolution amplifier; while, in the pulse case, there is the much narrower pulse, t_0, resulting in a transient response of the resolution amplifier. While the continuous-type spectrum is of major interest in pulsed RF, either type of display can be

The Sweeping-Signal Spectrum Analyzer

obtained by simply changing the resolution bandwidth of the spectrum analyzer. Table 5-1 details the major differences between the two types of display.

5.4. SENSITIVITY

Sensitivity is defined as a rating factor of the ability of the spectrum analyzer to display signals. Sensitivity is usually specified as the signal power that is equal to the analyzer noise power at a particular bandwidth; this is known as the "signal-plus-noise is equal to twice-noise" method. Spectrum analyzer noise level determines the sensitivity — less noise means better sensitivity. All amplifiers generate noise. Even an ideal amplifier would generate thermal noise, because of random current fluctuations in the input impedance. Thermal noise power can be computed from N = kTB, where

    k = Boltzmann's constant
      = $1.37 \cdot 10^{-23}$ watt-seconds/degree
    T = absolute temperature
      (measured from absolute zero, i.e., $-273°$ Celsius — usually assumed to be $290°$ Absolute)

The thermal noise power and, hence, the sensitivity is directly proportional to the bandwidth. A wider bandwidth means a poorer sensitivity. For example, an ideal amplifier having a 1 MHz bandwidth has a theoretical sensitivity of -114 dBm, while the same amplifier with a 100 kHz bandwidth has theoretical sensitivity of -124 dBm. Such calculations lead to the conclusion that for best sensitivity the spectrum analyzer should be operated at narrow resolution band-

Table 5-1

CW-Type Spectrum	Continuous-Type Spectrum
Lines on screen are Fourier spectral components	Lines on screen are repetition-rate lines
Line spacing depends on span setting and is independent of sweep time	Line spacing is determined by sweep time and is independent of span setting
Mathematical description is Fourier series	Mathematical description is Fourier integral
Resolution setting is $B <$ rep rate	Resolution setting is $B >$ rep rate
The CRT display shows the resolution amplifier steady-state response	The CRT display shows the resolution amplifier transient response
There is still energy in the circuit from previous pulses when the next pulse occurs	All the energy in the circuit from the previous pulse has decayed to zero when the next pulse occurs
Bandwidth-pulsewidth product is $Bt_0 > 1$	Bandwidth-pulsewidth product is $Bt_0 < 1$

widths. The conclusion is correct for discrete CW signals but not for pulse signals. This is because a discrete signal has a spectrum that, at least in theory, has zero frequency width. A reduction in bandwidth reduces the noise but should have no effect on the signal. On the other hand, a pulse signal, which generates a continuous dense type of spectrum, is affected by the resolution bandwidth. This is because the voltage of a continuous spectrum is defined on a per-unit-bandwidth basis, as is discussed in Chapter 3. This point is graphically illustrated in Figure 5-8. Here is shown the continuous (sin x)/x spectral distribution, typical of a rectangular pulse; energy is proportional to area squared, so the wider the resolution bandwidth the greater is the intercepted area and the larger the output. Eventually, when the bandwidth is so large that it intercepts most of the area, there is no further increase in output with an increase in bandwidth. However, the resolution bandwidth cannot be increased to this point without losing the fine details of the spectrum. As discussed previously, for proper definition of spectrum details the bandwidth should be about one-tenth of the inverse of the pulse width, $Bt_0 \leqslant 0.1$. Thus, for the same peak amplitude in time domain, a continuous spectrum for a pulsed signal will have a smaller amplitude than the discrete spectrum of a CW signal. This is intuitively apparent from the observation that equal peak amplitudes mean equal instantaneous power, which in one case is concentrated at one frequency and in the other case is distributed over a range of frequencies. Thus, there is a loss in sensitivity for pulsed signals as compared to CW signals.

The formula from which the loss in pulsed-signal sensitivity can be computed was first reported in Volume XI of the *Radiation Laboratory Series*.* This was later shown by Metcalf et al.† to be a simplified version of a more complicated expression. The simplified expression, which is sufficiently accurate for our purposes, is

$$\alpha \doteq \frac{3}{2} t_0 B$$

$$\alpha_{dB} = 20 \log_{10} \frac{3}{2} t_0 B \tag{5-9}$$

where t_0 is pulse width and B is the 3 dB bandwidth of a synchronously tuned gaussian set of amplifier stages. It can also be shown that the response of an ideal bandpass filter is $\alpha = t_0 B$. Here B is the impulse bandwidth. The previous expression is in full agreement with the second one for the special case of the

---

*Montgomery, "Techniques of Microwave Measurements," *Radiation Laboratory Series*, McGraw-Hill or Boston Technical Publishers, Vol. XI.

†Metcalf, Von Allmen, & Caprio, "Investigation of Spectrum Signature Instrumentation," *IEEE Trans.*, ECM-7, No. 2, June, 1965.

The Sweeping-Signal Spectrum Analyzer

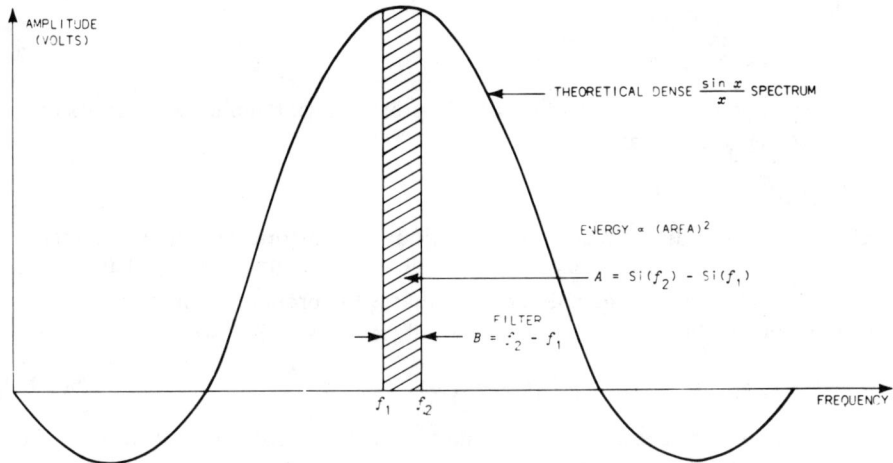

Figure 5-8 Continuous Spectrum Energy Distribution

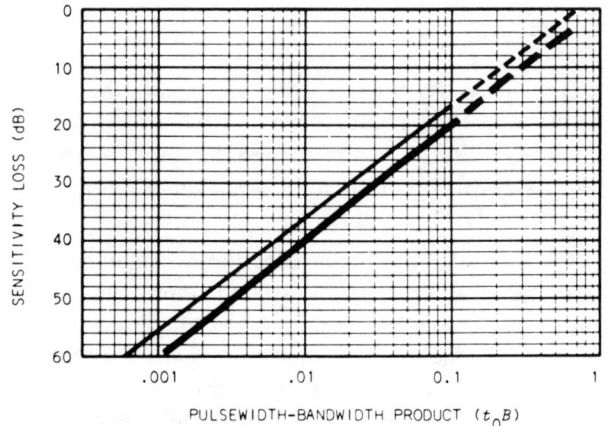

Figure 5-9 Loss in Sensitivity for Pulsed RF; Upper Trace $\alpha = 1.5\, t_0 B_3$; Lower Trace $\alpha = t_0 B_i$

gaussian filter. Sabaroff has shown that, for the gaussian filter, the ratio of impulse bandwidth (B_i) to the 3 dB bandwidth (B_3) is

$$\frac{B_i}{B_3} = \frac{2\pi}{4\sqrt{\pi \log \sqrt{2}}} = 1.5$$

The expression $\alpha = t_0 B_i$ is probably to be preferred as it applies to all types of filters, not just Gaussian.

5.5. CONVOLUTION

The convolution theorem is given in the Fourier transform section of Chapter 3. There it is indicated that the convolution of two time functions, $f_1(t)$ and $f_2(t)$, leads to a frequency domain description that is the product of the two frequency domain functions, $F_1(\omega)$ and $F_2(\omega)$ and vice versa.

$$\int_{-\infty}^{+\infty} f_1(\tau) f_2(t - \tau) \, d\tau \leftrightarrow F_1(\omega) F_2(\omega) \qquad (5\text{-}10)$$

Convolution is of special interest because it is a mathematical description of the relative translation of two functions, $f_1(t)$ and $f_2(t)$, where the variable τ indicates the relative movement of the functions. Convolution, therefore, describes the process taking place in the sweeping-signal spectrum analyzer, where $f_1(\tau)$ is the stationary filter and $f_2(t - \tau)$ is the moving signal. Since the integral of a unit impulse is unity, it follows that convolution with an impulse leads back to the original function.

$$\int_{-\infty}^{+\infty} f(\tau)\delta(t - \tau)d\tau = f(t) \qquad (5\text{-}11)$$

This means that the narrower the resolution bandwidth, the closer the displayed spectrum is to the theoretical input spectrum. Of course, there are constraints, such as sweep speed and span, on how narrow the resolution bandwidth can be made. Nevertheless, for theoretically ideal spectrum analysis, the resolution bandwidth has to be infinitely narrow.

The effect of the width of the sampling function, which in our case is determined by the resolution filter, is best illustrated graphically (see Figure 5-10). It is assumed there is a triangular function $f_2(t)$ sliding past a stationary rectangular sampling function $f_1(t)$, as is shown in Figure 5-10(A). The result is the graphs in Figure 5-10(C), which are the functions to be multiplied and integrated by the intermediate step, as is shown in Figure 5-10(B), which is just a change of variable.

Figure 5-10(D) illustrates the actual integration, which is just a determination of the area under the multiplied function as shown by the shaded area. Figure 5-10(E) is the final result, a distorted version of $f_2(t)$. As the width of the sampling function $f_1(t)$ gets narrower, the final result looks more and more like the

The Sweeping-Signal Spectrum Analyzer

input $f_2(t)$. In the limiting case, when the resolution filter has an infinitesimal bandwidth, the output function becomes identical to the input function. As a practical matter, it has been found that for a $(\sin x)/x$ spectral distribution, a filter bandwidth less than one-tenth of lobe width reproduces the original function with sufficient fidelity.

5.6. EXAMPLES

1. A spectrum analyzer is specified as having a certain sensitivity at a 100 kHz resolution bandwidth. The unit seems not to meet its specifications by 5 dB. The measurement conditions are:

Resolution bandwidth = 100 kHz
Span = 1 MHz/DIV or 10 MHz full screen
Sweeptime = 10 µs/DIV or 100 µs full screen

What is wrong? From Equation (5-3), the loss in sensitivity due to too fast a sweep rate is

$$\alpha = \left[1 + 0.195 \left(\frac{S}{TB^2}\right)^2\right]^{-1/4}$$

Substituting:

$$\alpha = \left[1 + 0.195 \left(\frac{10^7}{10^{-4} \cdot 10^{10}}\right)^2\right]^{-1/4} = (1 + 19.5)^{-1/4} = \frac{1}{2.13}$$

or a loss of $20 \log_{10} 2.13 \approx 6.6$ dB.

This accounts for the loss in sensitivity. Note that we cannot use precise numbers: Bandwidth B in Equation (5-3) is the 3 dB bandwidth, while the resolution bandwidth is frequently defined as the 6 dB bandwidth. However, because of errors caused by differences between the actual and assumed phase responses, it is impossible to get an accurate number no matter what bandwidth is used. All that the above calculation can tell us is that the discrepancy of 5 dB is not unreasonable.

2. It is desired to observe the spectrum of an FM signal. The approximate deviation is 100 kHz and the approximate FM rate is 5 kHz with a 100 MHz carrier. These numbers are well within the capability of many spectrum analyzers. However, the FM signal is initiated by an explosion and is expected to last no more than 1 ms. Is the measurement still possible?

We calculate the optimum resolution bandwidth which, from Equation (5-5) is:

$$B_o = 0.66 \sqrt{\frac{S}{T}}$$

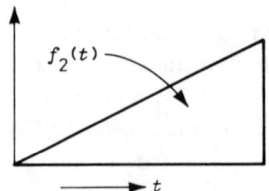 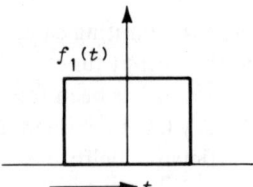

(A) Original Functions

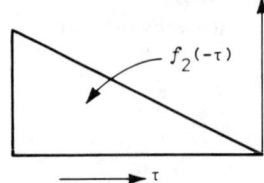 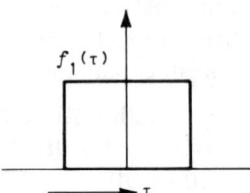

(B) Change of Variable from t to τ and $-\tau$

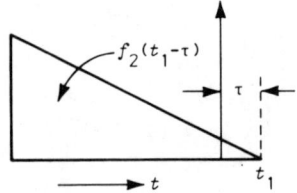 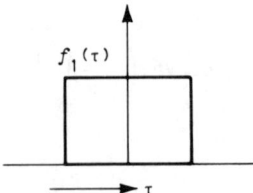

(C) Stationary Filter, Translating Signal Has Moved by t_1

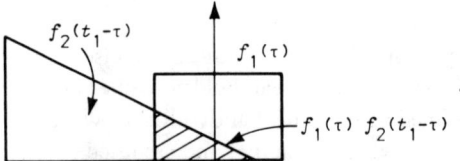 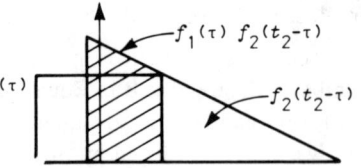

(D) Graphical Multiplication of Functions

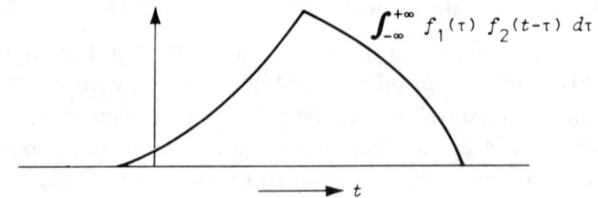

(E) Result of Convolution

Figure 5-10 Graphic Illustration of Convolution

$$B_o = 0.66 \sqrt{\frac{10^5}{10^{-3}}} = 6.6 \text{ kHz}$$

The actual resolution is $R_o = \sqrt{2} \cdot B_o \approx 9.3$ kHz.

Conclusion: The FM sidebands cannot be resolved.

3. It is desired to check a radar set operating at the rate of 10 pulses per second; what is the fastest reasonable sweep time for the analyzer? Assuming that we will observe the main lobe and one side lobe on either side, we need at least 20 rep-rate lines for appropriate definition of the spectrum envelope. Thus, we have to sweep at two seconds per screen diameter or slower.

4. The radar set in example 3 uses 1 μs pulses; what span should we use? The width of a side lobe is $1/t_0$ and the width of the main lobe is $2/t_0$. To observe the main lobe and two side lobes we need a full screen span of $4/t_0 = 4$ MHz or 400 kHz/DIV.

5. For proper spectral envelope definition, what is the widest permitted resolution bandwidth? The relationship is $t_0 B \leqslant 0.1$, resulting in $B = 0.1/t_0 = 100$ kHz.

6. What is the loss in sensitivity compared to CW under these conditions? The formula is $\alpha = (3/2) t_0 B$; therefore,

$$\alpha = \frac{3}{2} \, 10^{-6} \cdot 10^5 = \frac{3}{2} \, 10^{-1}$$

$$\alpha_{dB} = 20 \log_{10} \frac{3}{2} \, 10^{-1} = -16.5 \text{ dB}$$

or a loss in sensitivity of 16.5 dB.

7. Suppose the pulse rate is increased from 10 Hz to 200 kHz. Can a continuous spectral distribution still be obtained? The answer is no. We need a resolution bandwidth greater than the pulse repetition rate of 200 kHz in order to get a distributed spectrum display, but a resolution bandwidth that does not meet the requirements of $t_0 B \leqslant 0.1$ does not give adequate definition of the spectrum shape.

5.7. EXERCISES

5-1. Given the following spectrum analyzer settings:

Resolution bandwidth = 1 kHz
Span = 10 kHz/DIV
Sweeptime = 1 ms/DIV

a. What is the resolution (R) observed on the CRT?
b. What is the loss, if any, in CW sensitivity compared to a sweeptime of 1 s/DIV?

c. What is the optimum CW resolution bandwidth (B_0) setting for the given span and sweeptime?
d. What is the optimum resolution (R_0) for part c?

5-2. Given a spectrum analyzer having the following parameters:

Resolution bandwidth is continuously adjustable from 1 kHz to 1 MHz.
Span is continuously adjustable from 1 kHz/DIV to 100 MHz/DIV.
Sweeptime is continuously adjustable from 10 s/DIV to 1 μs/DIV.

a. What resolution bandwidth setting will result in the best sensitivity for a CW signal?
b. What will the sensitivity be in dBm assuming an ideal front-end amplifier?
c. What resolution setting will result in the best sensitivity for a 1 μs wide rectangular pulse train?
d. If the pulse repetition rate is 10 pulses per second, what is the fastest sweeptime that will provide an adequately detailed display of the spectrum?

Chapter 6
The Measurement Problem

6.1. INTRODUCTION

Many changes have occurred since the first sweeping spectrum analyzers were developed in the late 1930s. While much of the theory was developed during World War II, as described in Volume XI of the *Radiation Laboratory Series*,* the most significant changes in construction occurred in later years.

Today there are three broad classes of instruments, two which were introduced in the 1950s and 1960s, and 1970 - 1980, and those currently being introduced. The earliest instruments have no absolute calibration or spurious response immunity. Most are of swept IF design, though some swept front-end instruments can be placed in this category. An example is the Tektronix 491, first available in 1966 and withdrawn from the market in 1982. The second category provides absolute amplitude calibration and other improvements such as RF preselection to eliminate spurious responses or digital storage display. Most instruments available today are of this type. Examples are the Tektronix 7L12, first introduced in 1972, and the Tektronix 492, which was made available in 1979. The third class comprises the very latest instruments, which by means of microcomputers and other hardware, firmware, or software provide advanced measurement techniques such as multiple waveform storage, counter-type frequency accuracy, or signal manipulation and computation. The Tektronix 494P, introduced in 1984, is an example of this class.

6.2. TYPES OF MEASUREMENTS

There are four basic types of measurements that are desirable: absolute frequency, relative frequency, absolute amplitude, and relative amplitude. Not all spectrum analyzers have the sophisticated circuits that permit the absolute measurement of amplitude; however, most measurements do not require this. The most frequent need is for the determination of relative amplitude and relative frequency. Though there are, at most, four different types of measurements, some spectrum determinations can get quite complex. This is because in most instances it

---
Radiation Laboratory Series, McGraw-Hill or Boston Technical Publishers, Vol. XI.

is not just a matter of making a single relative-amplitude measurement or a relative-frequency measurement. Most measurement problems call for a succession of several measurements, the sequence of which is important; these measurements may each call for different spectrum analyzer control settings; and finally the measured data may have to be correlated or modified by computation before the final results are obtained. These aspects of measurements, which might be termed signal interpretation, are considered in subsequent chapters. In this chapter four basic parameters — relative frequency, absolute frequency, relative amplitude, absolute amplitude — will be considered.

6.3. MEASUREMENT LIMITATIONS

Given four basic measurements, namely relative and absolute amplitude and frequency, in combination with three classes of instrument — spurious prone uncalibrated, full amplitude calibration, and computer enhanced — the user might encounter up to 12 combinations. Most of these, however, are trivial and will not be discussed. The remainder of this chapter will be devoted to fundamental spectrum analyzer aspects, such as spurious responses for the swept IF design and the most common user traps and measurement problems inherent in current instruments.

6.4. SPURIOUS RESPONSE BASICS

6.4.1. Swept IF

All spectrum analyzers have a dial or other type of readout device that indicates the frequency that is supposed to correspond to the center of the CRT. Unfortunately, spectrum analyzers have spurious responses, so the readout does not always represent a true indication of an incoming signal. The major problem in absolute-frequency measurement is, therefore, to differentiate between the true response and the spurious responses. There are many types of spurious responses and these affect the spectrum analyzer to a different degree.

Three types of spurious responses affect the center-frequency readout capability. These are: IF feedthrough, image, and harmonic conversions. These have different effects and are identified in a different manner, depending on whether the spectrum analyzer is swept front-end or is of the swept IF variety. Consider the swept IF first.

Figure 6-1 is a block diagram of a basic swept IF system. A portion of the input frequency spectrum, having a maximum frequency width equal to the bandwidth of the first amplifier, is translated in frequency and applied to the second mixer where it is treated the same as an input to a swept front-end system.

In a properly designed system, the second conversion should not generate any spurious responses because the wideband amplifier controls the frequency width

The Measurement Problem

applied to the second mixer. Hence, our interest is in the first frequency conversion, which is described by Equation (6-1).

$$mf_{LO} \pm nf_{RF} = f_{mo}$$
$$nf_{RF} \pm mf_{LO} = f_{mo} \qquad (6\text{-}1)$$

where:

f_{LO} = local oscillator frequency
f_{RF} = signal input frequency
f_{mo} = mixer output frequency
m, n = positive integers including zero

The center frequency dial, or readout, is normally calibrated to indicate the local-oscillator frequency f_{LO}, or a harmonic of f_{LO}, in combination with the IF-amplifier center frequency f_0. From Equation (6-1) the possible combinations are:

$$f_d = mf_{LO} + f_0$$
$$f_d = mf_{LO} - f_0 \qquad (6\text{-}2)$$
$$f_d = f_0 - mf_{LO}$$

When the mixer output frequency (f_{mo}) is equal to the IF center frequency (f_0), there is a spectrum analyzer response. The dial frequency (f_d) may or may not be the same as the signal input frequency (f_{RF}) at that time. If the dial and signal frequencies are the same, there is a true response, while otherwise the response is spurious.

Consider, for example, the IF feedthrough. This occurs when the input signal frequency is equal to the IF-amplifier center frequency, thus $f_{RF} = f_0$. The local-oscillator frequency (f_{LO}) has nothing to do with this response; hence from Equation (6-2), it is clear that the dial indication (f_d) has no validity for this

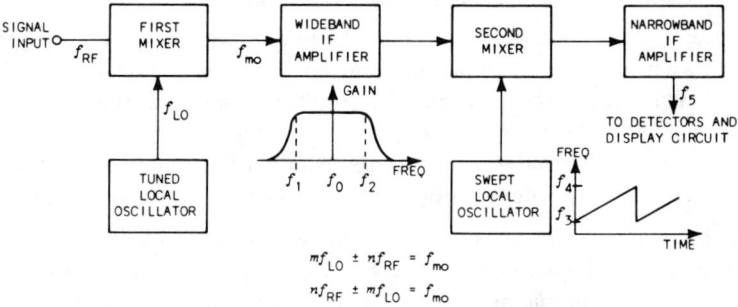

Figure 6-1 Swept IF Spectrum Analyzer, Basic Block Diagram

response. The IF-feedthrough spurious response can only occur for the narrow range in input frequencies that fall within the passband of the first IF amplifier. This frequency range is indicated as f_1 to f_2 in Figure 6-1. The IF-feedthrough spurious response is recognized by the fact that the setting of the RF-frequency tuning dial has no effect on it.

Another bothersome spurious response is the image. The image occurs when the signal frequency satisfies one of the mixer equations, but the dial is calibrated for one of the other two possible conversions. For example, suppose the dial is calibrated for $f_{LO} - f_d = f_0$. An input signal of frequency $f_{RF} = f_d$ would satisfy this equation and be a true response. However, an input signal whose frequency satisfies the equation $f_i - f_{LO} = f_0$, where $f_i \neq f_d$, will also appear on the screen. This second response is the image. The image frequency f_i and that of the true response, corresponding to the dial setting f_d, are separated by twice the IF center frequency f_0. This can be shown as follows:

$$\begin{array}{ll} f_{LO} - f_d = f_0 & \text{true response} \\ -f_{LO} + f_i = f_0 & \text{image response} \\ \hline f_i - f_d = 2f_0 & \end{array}$$

(6-3)

In a swept IF system, the image is recognized by the fact that the signal display will move across the screen in the opposite direction to the true response. This will be recognized from Equation (6-1), where for one conversion the output frequency increases with increasing local-oscillator frequency, while for the other conversion the output frequency decreases with increasing local-oscillator frequency. This is illustrated below.

$$\uparrow f_{LO} - f_d = f_{mo} \uparrow$$

$$f_i - \uparrow f_{LO} = f_{mo} \downarrow$$

where $\uparrow$ means increasing frequency and $\downarrow$ means decreasing frequency.

Of course, in order to determine which of the two responses is the image, one has to know which of the three possible conversions corresponds to the dial calibration. When it is discovered that the on-screen response is the image, it is necessary that the dial setting be changed by twice the wideband IF center frequency in order to obtain the true response. Whether the dial numbers have to be increased or decreased depends on which conversion the dial is calibrated for.

The last of the spurious responses that can cause ambiguity in absolute-frequency measurements is the harmonic conversion response. This is due to the signal combining with a harmonic of the local oscillator to produce an IF frequency output at a dial setting, which does not correspond to the signal frequency.

Not all harmonic conversions are spurious. Many spectrum analyzers utilize har-

The Measurement Problem

monic conversions to increase frequency coverage. Harmonic spurious responses are identified by the rate of movement across the CRT as a function of RF center frequency tuning. This is because the rate at which the mixer output frequency (f_{mo}) changes is determined by $m_t f_{LO}$, while the number on the dial changes at the rate of $m_d f_{LO}$. Unless the two harmonic numbers, m_d and m_t, are the same, the rate of frequency change on the dial will not agree with the rate of frequency change on the CRT. In practice, the test is to make a frequency difference measurement using the RF center frequency dial and to compare the numbers obtained with a frequency difference determined from the span. If the response on the screen is a harmonic spurious response, the two numbers will disagree by an integer fraction such as 1/2, 2/3, 4/3, etc. If, for example, the dial is calibrated for $m_d = 1$, while the signal frequency is such that to get the IF center frequency (f_0) it must combine with $m_t = 2$, tuning the signal across the full span of the screen will result in a dial number change which is one-half that of the frequency span.

6.4.2. Swept Front-End

So far, we have considered how to verify the center frequency reading for a swept IF system. Similar problems exist, although not to the same extent, for the basic swept front-end system. Figure 6-2 is a basic block diagram of a swept front-end unit. The spurious response problem is much alleviated in this system because the first IF amplifier, being of the narrowband variety, can be constructed at a much higher frequency than the wideband unit in a swept IF system. A higher IF amplifier frequency means greater frequency separation between the true response and the spurious responses and, hence, fewer difficulties. Of course, one pays for this with greater sweeper-system complexity – hence, more weight, size, and cost. While this type of system has the same type of spurious responses as the swept IF unit, the appearance and identification of these responses are different and sometimes more difficult. Consider these spurious responses in turn.

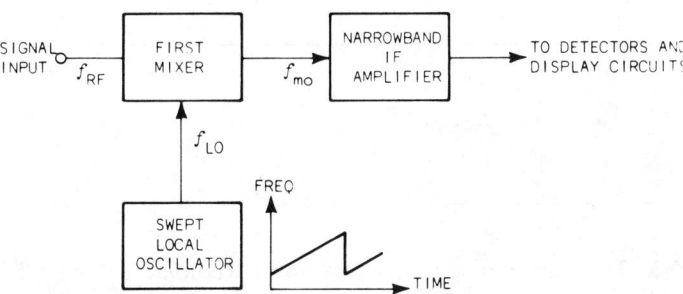

Figure 6-2 Swept Front-End Spectrum Analyzer, Basic Block Diagram

The IF feedthrough response is caused by an input signal whose frequency is within the passband of the first IF amplifier, and which does not enter into a conversion or mixing with the first local oscillator. In the swept front-end unit, the first local oscillator is the sweeping local oscillator. Hence, the IF feedthrough is not a swept signal. This means that a continuous-wave input is not converted into its frequency domain equivalent, i.e., a narrow pulse. The continuous-wave input exists at all times within the passband of the IF amplifier, causing the whole baseline to deflect or rise. The IF feedthrough spurious response is, therefore, recognized by a shift in the baseline level.

As in the swept-IF system, the image frequency is separated from the true response frequency by twice the IF amplifier center frequency. However, unlike the swept-IF system, the image response does not move on the screen in an opposite direction to the true response as the center-frequency dial is tuned.

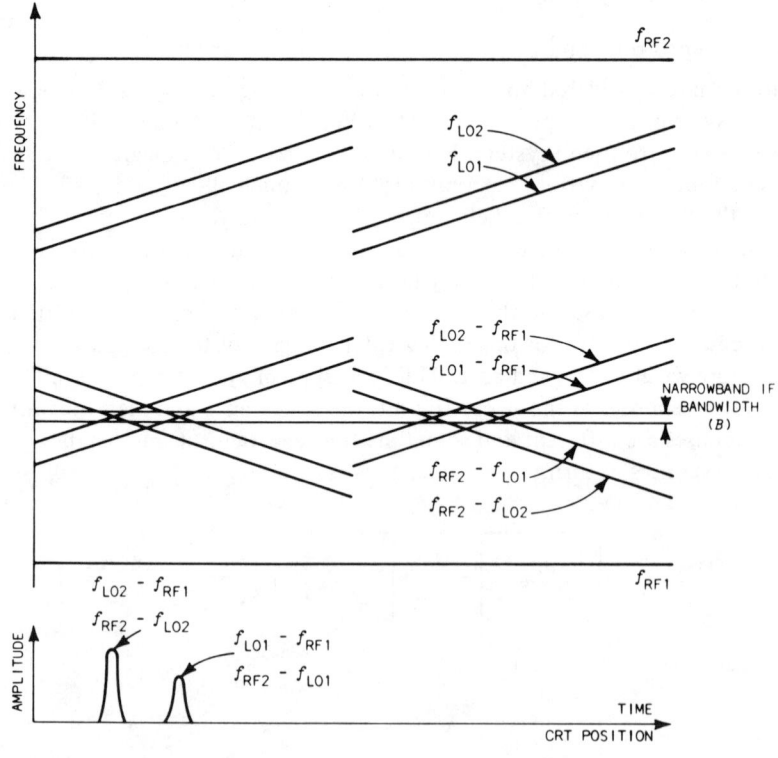

Figure 6-3 Time/Frequency Diagram of Image Response Tuning in Swept Front-End Spectrum Analyzer

The Measurement Problem

This point is demonstrated in the time/frequency diagram, Figure 6-3. Two CW signals are applied to the spectrum analyzer at frequencies f_{RF1} and f_{RF2}, respectively. One uses the $f_{LO} - f_{RF1} = f_{mo}$ response, while the other corresponds to the $f_{RF2} - f_{LO} = f_{mo}$ response. One of these responses is the image, while the other is the true response — for our purposes it does not matter which is which. As the sweeping local-oscillator center frequency is changed, the frequency sawtooth moves from the curve labeled f_{LO1} to the one labeled f_{LO2}.

The corresponding mixer outputs are $f_{LO1} - f_{RF1}$ and $f_{LO2} - f_{RF1}$ for one response and $f_{RF2} - f_{LO1}$ and $f_{RF2} - f_{LO2}$ for the other response. From Figure 6-3 it will be observed that regardless of which conversion is utilized the response on the CRT screen moves the same way. Hence, there is no way to distinguish the image from the true response simply by operating the spectrum analyzer controls in a swept front-end system.

6.5. MORE ON TRUE SIGNAL IDENTIFICATION

The vast majority of modern spectrum analyzers are of swept front-end design, and most of these include a sweeping filter preselector prior to the first mixer. The preselector tracks the instantaneous input signal frequency by sweeping in synchronism with the local oscillator or a desired local oscillator harmonic but offset by the first IF frequency. Consequently, the three most troublesome spurious responses discussed previously, image, IF feedthrough, and harmonic conversion, are virtually eliminated at microwave frequencies up to about 20 GHz. These types of spurious responses are still a problem at higher frequencies and for the less sophisticated, unpreselected, spectrum analyzers. However, rather than manually comparing tuning rates or direction against frequency span, modern instruments rely on a frequency identifier mode.

Frequency identification is still based on the techniques previously described, but the process is more or less automated. Here the sweeping/tuning first local oscillator is offset in frequency a predetermined amount on alternate sweeps while the gain is reduced. The result is that a real signal will shift sideways a fixed distance, usually two divisions, and drop in amplitude on alternate sweeps. Spurious signals will shift the wrong distance or in the wrong direction. A more sophisticated procedure is to simultaneously shift another oscillator in an offsetting direction so that a real signal will not move sideways but drop in amplitude on alternate sweeps. Spurious signals will move sideways as well as drop in amplitude. Figures 6-4(A) and 6-4(B) illustrate the true signal identification mode.

The above procedure works well but is still lacking under some conditions. The identify mode can only be actuated at a pre-set span position, otherwise the sideways shift, or lack of shift, will not be calibrated. Bandwidth and oscillator

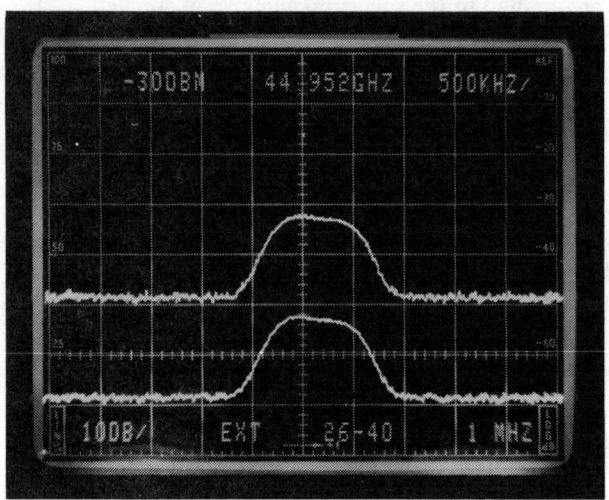

(A) Real Signal

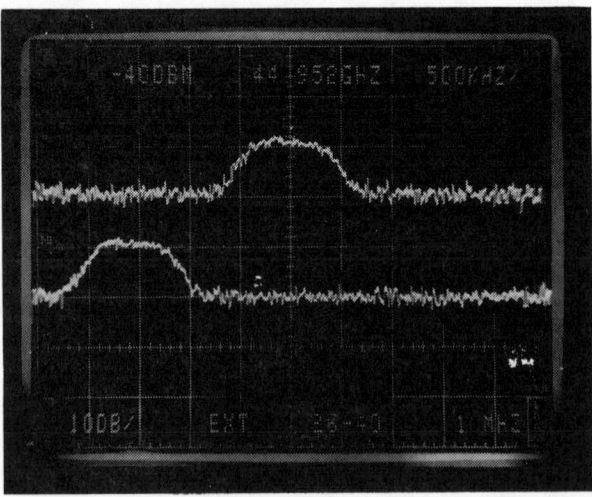

(B) False Signal

Figure 6-4 Signal Identify Mode for Ordinary Spectrum Analyzer Operates Only at 500 kHz/DIV

The Measurement Problem 115

tuning conditions limit the setting to a few hundred kHz/DIV. Furthermore, the setting accuracy with which local oscillator frequency can be shifted is at best kilohertzes, if not worse. This means that it is difficult to judge the position of unstable signals that move about 100 or more kilohertz, a common condition at millimeter wave frequencies. Wideband, non-CW signals are difficult to identify. Also the slight shift due to a spurious signal at a high conversion harmonic (e.g., X 21) as opposed to no shift for a true signal at an adjacent high harmonic conversion (e.g., X 20) is virtually unobservable. The user must resort to some clever guessing, signal adjustment, or additional equipment to help with the identification. For broadband signals it is best, for instance, to center a relatively narrow feature of the spectrum as opposed to the true center of the spectrum. For a pulsed RF sin x/x spectrum, it is better to key off a null rather than the broad mainlobe. At millimeter wave frequencies one may have to resort to a wave meter. These problems are almost completely eliminated in the newest, computer-enhanced instruments.

The most modern instruments, such as the Tektronix 494, provide extremely high frequency accuracy. On the order of a few hertz at a few megahertz, and kilohertz at hundreds of gigahertz. This kind of absolute accuracy permits large, very accurate oscillator frequency shifts. This provides frequency identification at spans about 10 times that of previous spectrum analyzers — 5 MHz/DIV versus 500 kHz/DIV, for example. Furthermore, internal computing power permits identification of operating mode setting and readjustment of computing algorithm so that signal identification is no longer limited to just one frequency span. With these instruments it is virtually impossible to make a mistake as to the true input signal frequency. Such a display is illustrated in Figure 6-5.

6.6. OTHER SPURIOUS RESPONSES

The modern spectrum analyzer has virtually eliminated major spurious responses, and spurious displays are now fairly easy to identify when these do occur. However, several other types of "not true" displays are possible in addition to those previously discussed. The most important are displays not due to any input signal known as residual responses, and spurious displays caused by nonlinear distortion in the IF chain. Residual responses occur when oscillators internal to the spectrum analyzer, or combinations of such oscillators, fall within the amplifier chain. The problem with residual responses has actually gotten worse and not better. This is because modern instruments have greater stability and better resolution and sensitivity than previous instruments, hence smaller residual responses can be observed. In addition, the indirect and direct synthesis chains and computing clocks of modern instruments create a larger group of internally generated signals that can cause residual responses. Hence, even though the residual response level has actually improved as a result of greater care in design, concern with residual responses has also increased.

116 Modern Spectrum Analyzer Theory and Applications

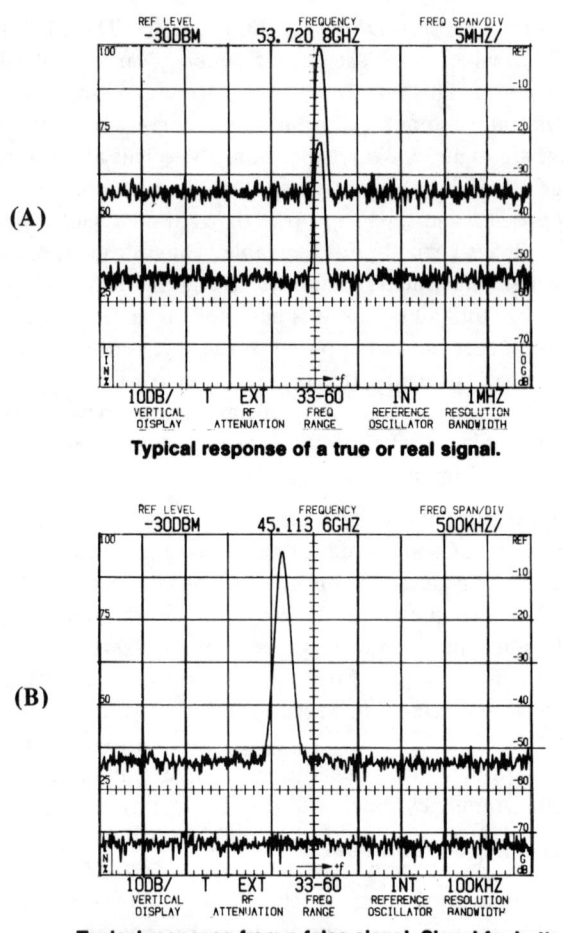

Figure 6-5 Signal Identify Mode for Computer Enhanced Spectrum Analyzer Operates at Wide 5 MHz/DIV Span

The residual response can fool even the most experienced user because some of these can appear to be real signals even when checked with the signal identifier. Fortunately, this is a very rare occurrence because residual responses are few in number and of low amplitude, under −100 dBm. These will occur at very specific frequencies associated with internal oscillators or combinations of oscillators. For instance, the 2 × 2 response associated with the second harmonics of the first two oscillators is a potential residual response. Thus, suppose the

The Measurement Problem 117

first LO tunes at 2-4 GHz, the first IF is at 2 GHz, the second LO is at 2.1 GHz and the second IF is at 0.1 GHz. The spectrum analyzer is made to tune from zero to 2 GHz [(2-4 GHz LO) − (0-2 GHz signal) = 2 GHz IF]. Any first LO at 2.05 GHz that gets into the second mixer, either by radiation or due to inadequate selectivity in the 2 GHz IF, will combine with the second LO, thusly $2 \times 2.1 - 2 \times 2.05 = 0.1$ GHz, the second IF. The 2×2 response will always occur at an equivalent input frequency equal to one-half the second IF, which in this example is 50 MHz. The only certain way of identifying a residual response is to disconnect the input signal and to see what remains.

A major class of spurious responses is introduced as a result of nonlinear distortion in the signal path. Harmonics and intermodulation (IM) are good examples of this class of undesired responses. Instrument performance is specified under the heading of harmonic distortion, intermodulation, and/or dynamic range. These specifications are discussed in Chapter 11.

6.7. EXERCISES

6-1. The spurious response relationships are very complicated for the swept-IF type design. Modern instruments no longer use this arrangement. However, an analysis of such a spurious response situation can be most instructive. A solution has been provided as an aid to the student.

Refer to Figure 6-6, which shows a CRT display involving five signals. Which of the responses are true, which are spurious and what are the input frequencies?

For a swept-IF system with local-oscillator frequency above signal frequency, a true response will move from left to right as the dial frequency is increased. Response a is the only one that moves from left to right with increasing dial frequency, all the other responses either move from right to left or stand still. Hence, a is either a true response or a harmonic conversion, and all the other responses are spurious. To determine whether a is a true response, note that, at a span of 5 MHz/DIV, a has moved four divisions as the center frequency dial has moved from 300 MHz to 320 MHz. Since $320 - 300 = 4 \cdot 5$, the span and the tuning dial agree; therefore, a is a true response. The input frequency causing this response is 305 MHz because: When the center of the CRT corresponds to 300 MHz, a is one division, or 5 MHz, to the left of center — for a converted signal, left means higher frequency, hence, 305 MHz.

Now take the spurious responses in alphabetical order. Response b is in the center of the screen and has not moved. It is an IF-amplifier feed-through response, which makes it 200 MHz. Response c is also an IF feed-through because it has not moved. This response is two divisions away from center so the signal is $2 \cdot 5 = 10$ MHz from 200 MHz. This signal is at 210 MHz because, for an unconverted signal, the screen represents increasing frequency going from left to right.

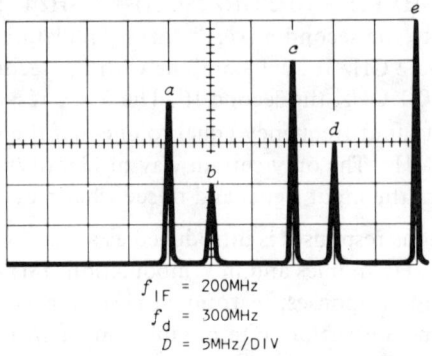

(A) f_{IF} = 200 MHz; f_d = 300 MHz; S = 5 MHz/DIV

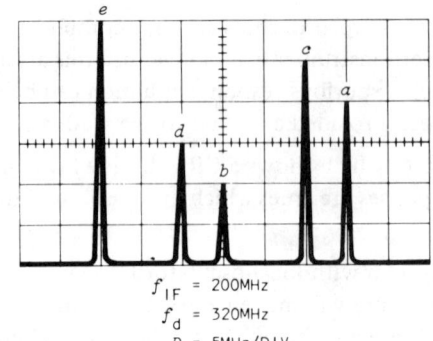

(B) f_{IF} = 200 MHz; f_d = 320 MHz; S = 5 MHz/DIV

Figure 6-6 Complex Spurious Display

Consider now response d. This response moves in the wrong direction, so it is an image. Response d moves 20 MHz when the dial moves 20 MHz, so it represents a prime conversion. Hence, when the dial reads 300 MHz, the local-oscillator frequency is $f_{LO} = f_d + f_{IF} = 300 + 200 = 500$ MHz. For the image, $f_{RF} - f_{LC} = f_{IF}$, $f_{RF} = 200 + 500 = 700$ MHz for the center of the screen. But, when the dial reads 300 MHz, response d is three divisions to the right of center. While, for a true response, the screen represents increasing frequency left to right; for the image, it is right to left. Hence response d is 3 · 5 = 15 MHz above screen center, or 700 + 15 = 715 MHz.

Finally, response e is an image since it moves from right to left with increasing dial reading. This image is, however, not a fundamental conversion but a second harmonic conversion, since the response moves eight divisions, representing 40 MHz according to the span, when the dial has moved only 20 MHz. Again, when the dial is 300 MHz, the local oscillator is 500 MHz. The second harmonic-image spurious response is based on the conversion $f_{RF} - 2f_{LO} = f_{IF}$. Hence, $f_{RF} = 200 + 2 \cdot 500 = 1200$ MHz for the center of the screen. The response is, however, five divisions to the right of center, which for an image means higher in frequency. Hence, at 5 MHz/DIV, response e due to a signal at a frequency of $1200 + 5 \cdot 5 = 1225$ MHz.

6-2. Assuming a nonlinear mixer characterized by the transfer function $i = a_1 e + a_2 e^2 + a_3 e^3$, where i is output current and e is input voltage, show that harmonic and intermodulation products are developed for a voltage input e = A sin s + B sin o, where s is signal frequency and o is local oscillator frequency.

6-3. Identify (name) the following spurious responses:
a. Does not move with main tuning dial.
b. Moves at the same rate but opposite direction as real signal.
c. Appears at twice the frequency of the input signal and changes amplitude more than inserted RF attenuation.

Chapter 7
Amplitude Modulation

7.1. BASIC RELATIONSHIPS

The following fundamental relationships apply to normal double-sideband amplitude-modulation (AM) measurements:

1. In the time domain, the percent modulation is computed from

$$K = \frac{E_{max}}{E_{min}} = \frac{(1+m)A}{(1-m)A} \quad m = \frac{K-1}{K+1} = \frac{E_{max} - E_{min}}{E_{max} + E_{min}} \quad (7\text{-}1)$$

   where E_{max} and E_{min} are as in Figure 7-1.

2. In the frequency domain, the percent modulation is computed from

$$m = \frac{2A_s}{A_c} \times 100 \quad (7\text{-}2)$$

   where A_s and A_c are per Figure 7-2; A_s/A_c is a voltage ratio.

3. The spectrum consists of a carrier and two equal amplitude sidebands for each modulating frequency. The sideband spacing with respect to the carrier frequency is equal to the modulating frequency.

4. In AM, the carrier spectral component is of constant amplitude regardless of the degree of modulation.

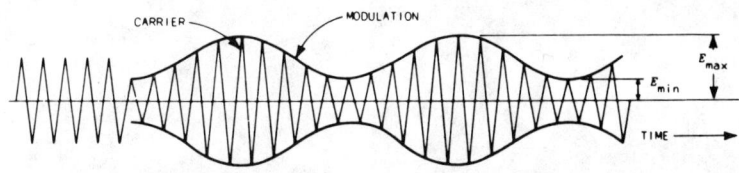

Figure 7-1 Time Domain AM

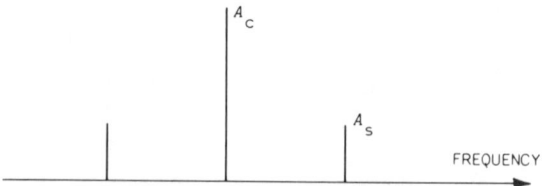

Figure 7-2 Frequency Domain AM

7.2. MEASURING AM

Figures 7-3 to 7-5 illustrate normal AM measurements.

Figure 7-3(A) shows the time domain (oscilloscope) appearance of a 10 MHz carrier, modulated by a 8 kHz signal. At a sweep time of 100 μs/DIV, the period of the modulating wave is about 125 μs or a frequency of 8 kHz. The 10 MHz carrier frequency could also be determined by operating at a faster sweep. To determine the percentage modulation, observe that

$$K = \frac{E_{max}}{E_{min}} = \frac{3}{1} = 3 \quad \text{and} \quad m = \frac{K-1}{K+1} = \frac{2}{4} \cong 50\%$$

Figure 7-3(B) shows the same signal in the frequency domain (spectrum analyzer). From the center-frequency setting, observe that the carrier is at 10 MHz. The modulating frequency is 8 kHz, since the sidebands are spaced about 1.6 divisions from the carrier at a span of 5 kHz/DIV. At 2 dB/DIV, the carrier is 12 dB larger than the sidebands. This is a voltage ratio of 4. Hence, the percentage modulation is m = $2A_s/A_c$ = 2/4 ≅ 50%.

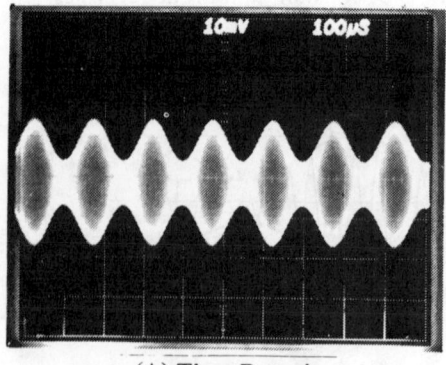

(A) Time Domain

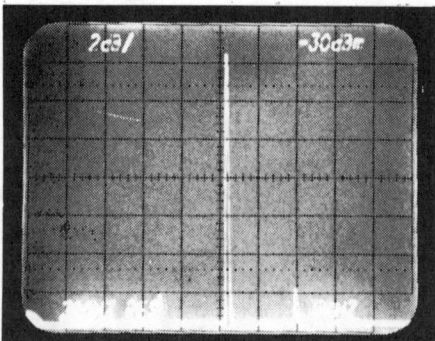

(B) Frequency Domain

Figure 7-3 Single Tone AM

Amplitude Modulation 123

For relatively high levels of modulation (e.g., over 10%), the oscilloscope and spectrum analyzer yield basically the same information. However, for small levels of modulation, the spectrum analyzer is definitely superior, as is illustrated in Figure 7-4. Both measurements indicate the frequencies, but the spectrum analyzer determination of percentage modulation is much easier. Here, the sidebands are 42 dB down. Since 42 dB represents a voltage ratio of about 0.008, the percentage modulation is about 1.6%.

Multitone modulation is another case where data is easier to obtain from the spectrum. This is illustrated in Figure 7-5. Figure 7-5(A) is the time domain appearance of a multitone AM wave. Not only is it difficult to ascertain the degree of modulation, but it is virtually impossible to determine the frequencies involved. Figure 7-5(B) is the frequency domain appearance of the same waveform. From this it is apparent that there are two modulating frequencies, about 9 kHz and 6 kHz. The percentage modulation at 6 kHz is about $2(1.5)/6.8 \cong 45\%$, and at 9 kHz, it is about $2(.6)/6.8 \cong 18\%$.

7.3. OTHER FORMS OF AM

A form of AM that saves power is double-sideband suppressed-carrier modulation. Here the interest centers on the degree of carrier suppression rather than on the degree of modulation. The amplitude of the carrier is measured relative to the sidebands, usually with the transmitter operating at the rated peak envelope power (PEP). Figure 7-6 is a spectrum analyzer display of a double-sideband suppressed-carrier amplitude-modulated wave. At 10 dB/DIV, the carrier amplitude is 35 dB below that of the sidebands. The time-domain appearance of double-sideband suppressed-carrier AM is similar to that of standard AM at

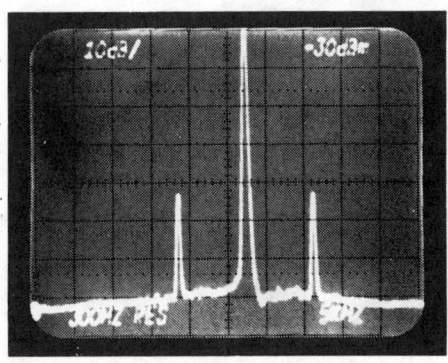
(A) Frequency Domain

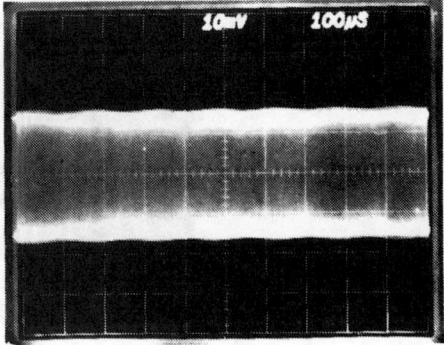
(B) Time Domain

Figure 7-4 Low Level Single Tone AM

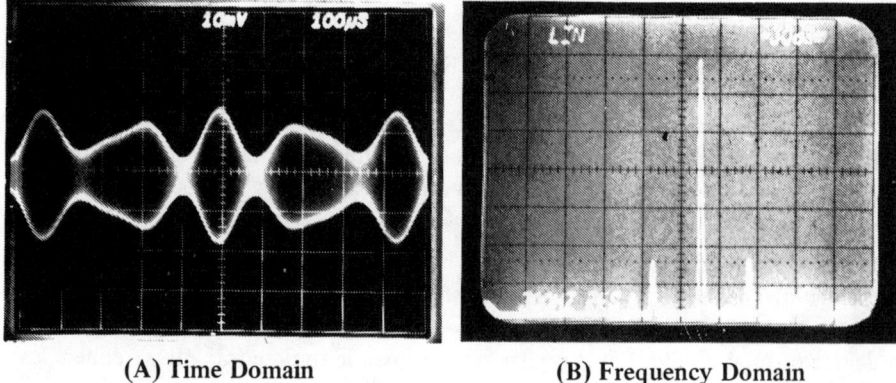

(A) Time Domain (B) Frequency Domain

Figure 7-5 Multitone AM

100% modulation. With some care these can, however, be distinguished. Figure 7-7 illustrates the difference in the time-domain appearance of these two forms of modulation.

Another frequently utilized form of AM is single sideband. Here only one sideband is transmitted, while the other sideband and the carrier are suppressed. This saves power and conserves frequency space. The sideband suppression is measured in the same manner as the carrier suppression ratio previously discussed.

Besides measuring carrier or sideband suppression, the spectrum analyzer can also be used to determine the peak envelope power. The PEP cannot be meas-

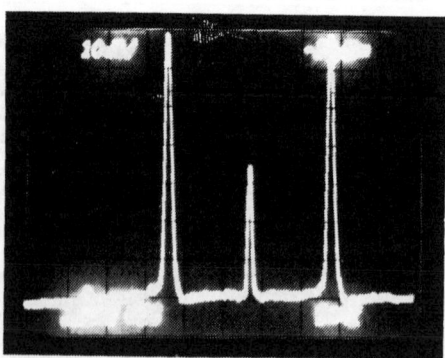

Figure 7-6 Double Sideband Suppressed Carrier AM in Frequency Domain

Amplitude Modulation

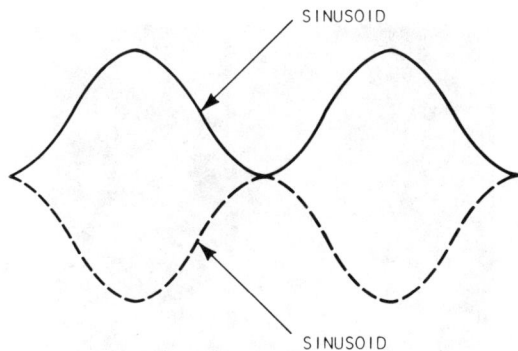

(A) Ordinary AM 100% Modulation, Sinusoids Do Not Intersect

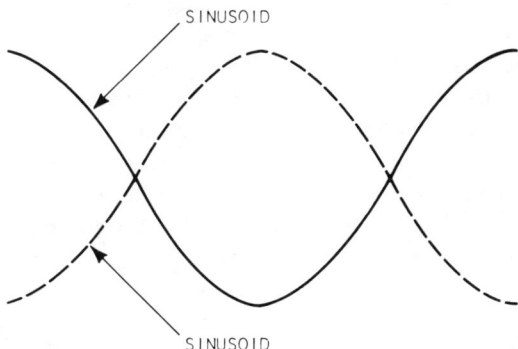

(B) Double Sideband Suppressed Carrier AM, Sinusoids Intersect

Figure 7-7 Time Domain Difference Between Ordinary 100% AM and Double-Sideband Suppressed-Carrier AM

ured with a standard power meter since these measure the true rms power. A peak power meter that responds to the peak of the waveform will do the job. The PEP can also be computed by determining the peak rms voltage and computing the power from that. Thus, for Figure 7-6, assuming a 50 Ω system, each of the sidebands is -30 dBm rms, which is equivalent to 7 mV. The carrier at -65 dBm is about 0.13 mV rms. The peak voltage is 14.13 mV rms. And the PEP is $(14.13 \times 10^{-3})^2/50 = 4 \times 10^{-3}$ mW or -24 dBm. The average power is the sum of the individual rms values, which is about -27 dBm.

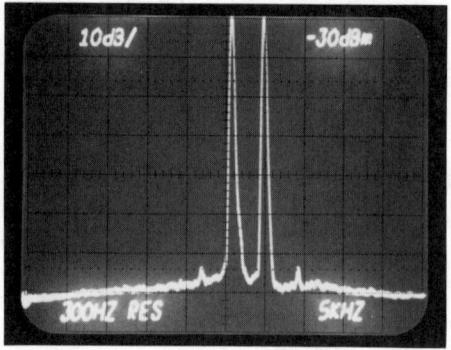

Figure 7-8 Two-Tone Intermodulation Measurement

A measurement that is of particular interest in single sideband is that of intermodulation distortion (IM). The measurement is performed by modulating the single sideband transmitter with two audio tones and checking the output with a spectrum analyzer for additional sidebands. Here a spectrum analyzer having good IM immunity is essential. Transmitters commonly have over 40 dB of IM immunity and some are as good as 60 dB down. Figure 7-8 shows a two-tone IM measurement. The third order intermodulation sidebands are 62 dB below the tone level. This spectrum analyzer has over 70 dB of IM rejection; hence, the 62 dB number represents the transmitter output.

7.4. EXERCISES

7-1. Refer to Figure 7-9. A single side-band transmitter is modulated at a 2 kHz rate.
   a. Is the small signal to the right the suppressed second sideband or the suppressed carrier?
   b. What is the degree of suppression?
7-2. Refer to Figure 7-10. Given a transmitter operating in the standard double sideband AM mode, the additional sidebands are due to distortion in the modulating waveform.
   a. What is the modulating frequency?
   b. What is the percentage modulation?
   c. What is the percent distortion of the modulating waveform?
7-3. Refer to Figure 7-11. Given a single sideband transmitter modulated by three tones,
   a. If the small signal 1.5 divisions to the left of the tones is the carrier, what are the frequencies of the modulating tones?
   b. What is the PEP?

Amplitude Modulation

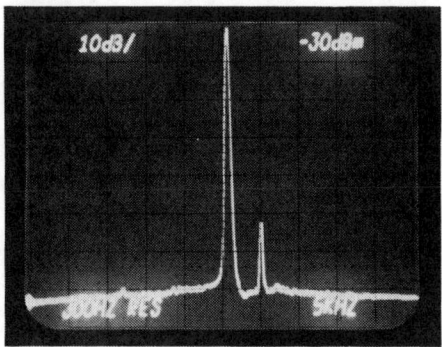

Figure 7-9 Exercise 7-1

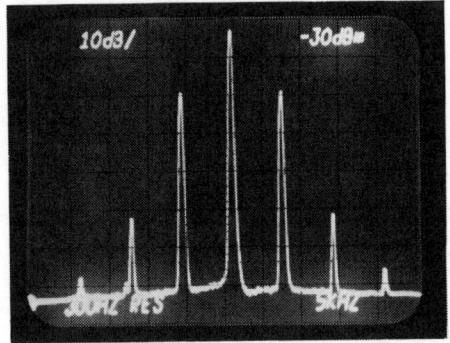

Figure 7-10 Exercise 7-2

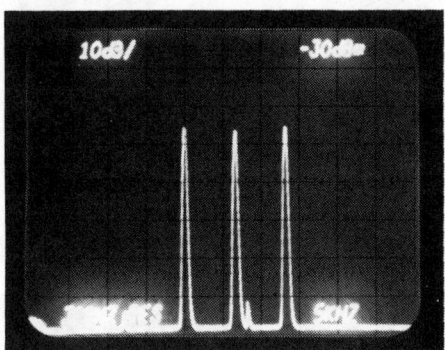

Figure 7-11 Exercise 7-3

Chapter 8
Frequency Modulation

8.1. BASIC RELATIONSHIPS

The following basic relationships apply to frequency modulation (FM).

1. Frequency modulation is a constant power process. The power of the modulated wave does not change as the degree of modulation changes.
2. The frequency domain representation of an FM wave consists of a carrier and sidebands spaced in frequency around the carrier. The spacing between frequency components is equal to the modulating frequency, f.
3. Theoretically, the FM wave contains an infinite number of sidebands. The sideband energy, however, falls off very rapidly outside the peak frequency deviation. Deviation is measured with respect to the carrier frequency.
4. The amplitudes of the various frequency components, including the carrier component, change as the deviation changes. This is a consequence of the requirement that the total power remain constant regardless of the deviation.
5. The relative amplitudes of the frequency components are in the same relationship as the relative amplitudes of Bessel functions of the first kind. Bessel functions of the first kind are designated by the letter J. The complete characterization of the frequency component amplitudes is $J_p(\Delta F/f)$, where p is called the order and represents the frequency component number (p = 0 for the carrier, p = 1 for the first sideband, etc.), and $\Delta F/f$ is called the argument and represents the modulation index. The modulation index (sometimes designated as m, t, or β) is defined as the ratio: peak frequency deviation ΔF divided by the modulating frequency f.
6. Bessel functions are the solution to a certain differential equation, just as the standard trigonometric functions (sine and cosine) are the solution to a specific differential equation. Graphs and tables of Bessel functions of the first kind are readily available. Figure 8-1 is such a graph.
7. The information of interest in FM is: the carrier frequency (F), the modulating frequency (f), and the deviation (ΔF). The carrier frequency F is obtained by reading the spectrum analyzer center-frequency dial and the modulating frequency f is obtained by calculating the frequency spacing between

two adjacent components by use of the calibrated span. The deviation (ΔF) can, however, not be determined directly. First, one obtains the modulation index from which the deviation is then calculated. Most of what follows pertains to the calculation of the deviation. A detailed theoretical discussion of FM will be found in Chapter 4.

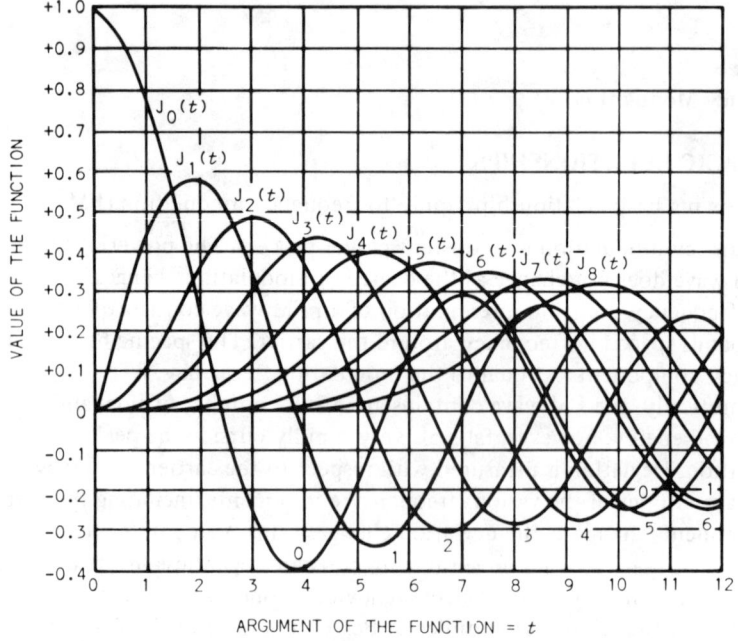

(A) Bessel Functions for the First Eight Orders

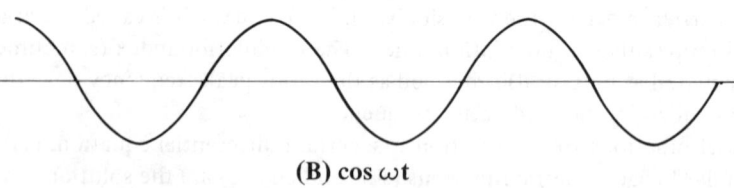

(B) cos ωt

Figure 8-1 Graph of Sinusoid and Bessel Functions

Methods of determining the deviation differ, depending on the modulation index. Techniques that work well at fractional indices (e.g., $\Delta F/f \ll 1$) will not yield any useful data for relatively large indices (e.g., $\Delta F/f > 10$). While no standard designation exists, it is convenient, for the purposes of this book, to separate FM into three deviation regions: narrowband, wideband, and ultrawideband FM. It is emphasized that these are not standard designations and should not be confused with similar names found elsewhere. For the purposes of discussion, narrowband FM means a modulation index less than unity, wideband FM will refer to modulation indices from about 1 to about 10, and ultrawideband FM will refer to modulation indices greater than 10.

8.2. NARROWBAND FM

Because sideband energy falls off very rapidly outside the peak frequency deviation, narrowband FM is characterized by only two significant sidebands. This is especially true at modulation indices less than about 0.5, where it is difficult to distinguish, from a spectrum analyzer display, whether the signal is AM or FM. If one knows that the signal is FM, one can proceed directly to the problem of determining the various modulation parameters. If, however, the basic nature of the signal is not known, it is necessary first to determine whether it is AM or FM. With spectrum analyzers that have a zero span position, it is possible to do this for modulation rates up to about one-half the widest resolution bandwidth of the spectrum analyzer.

AM is distinguished from FM by the basic difference in the methods used in detecting them. Amplitude modulation can be detected by an ordinary diode peak detector, whereas to detect frequency modulation, it is necessary to use a discriminator. While spectrum analyzers do not usually contain a discriminator for detecting FM, this can be done by slope detecting the FM signal on the skirts or slopes of the resolution amplifier curve. This is illustrated in Figure 8-2. The horizontal lines represent the successive center frequencies of the frequency-modulated signal as it is tuned through the range of the resolution amplifier resonance curve by means of the spectrum analyzer fine center-frequency control. With the spectrum analyzer set for 0 Hz/DIV span (i.e., not sweeping), the detected modulating signal appears directly on the CRT. The amplitude of this detected signal depends on the slope of the resonance curve and the deviation of the frequency-modulated signal. At positions A and E, the slope of the resonance curve is small, resulting in very little signal output. At position B, the output voltage is $(+V_1) + (+V_2) = (V_1 + V_2)$. At position D, the output voltage is $(-V_5) + (-V_6) = -(V_5 + V_6)$, where the minus sign denotes the change in slope between positions B and D. At position C the output is $V_3 - V_4$, which is very small if the curve is reasonably symmetrical. The result is, for an FM signal, two positions of maximum output voltage occurring around the middle of the resonance curve. The edges and peak of the resonance curve yield very little output. Figure 8-3 shows the actual spectrum analyzer display for such a measurement. For amplitude modulation, there is only one position of maximum output — at the peak of the resonance curve. Having established, from prior knowledge or through the above procedure, that the signal in question is narrowband FM, it is now possible to proceed with the basic measurement.

132 Modern Spectrum Analyzer Theory and Applications

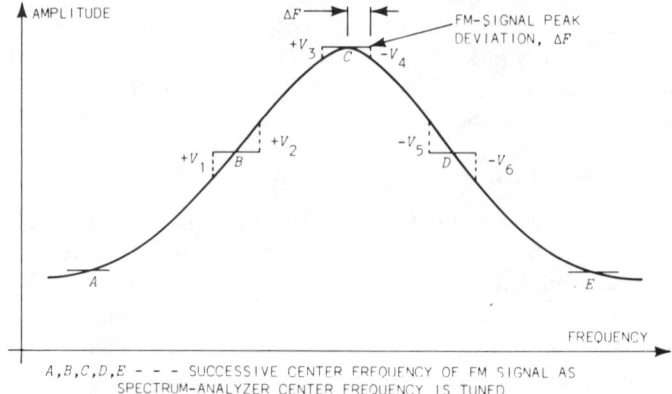

Figure 8-2 Slope Detecting an FM Signal

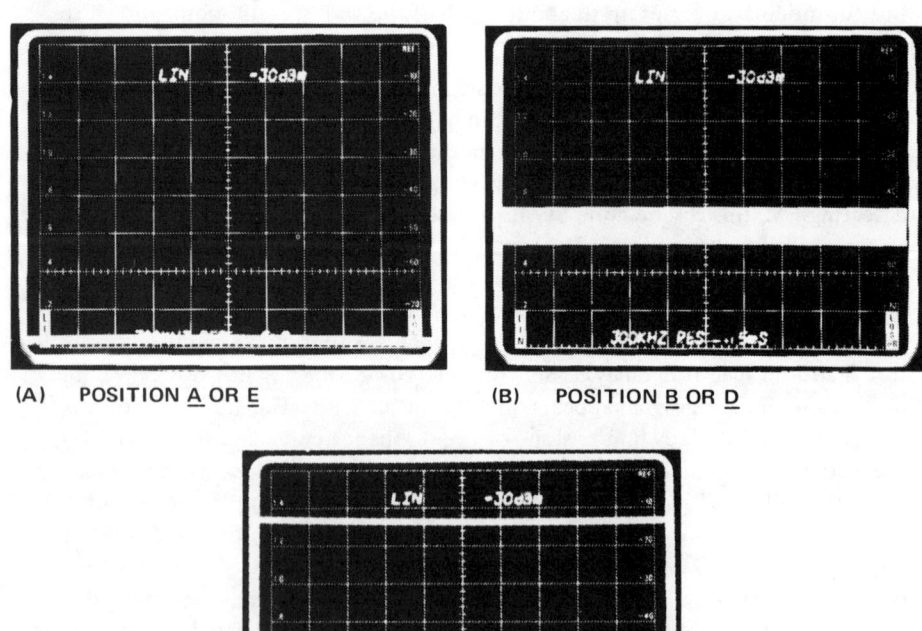

Figure 8-3 Detecting FM with a Spectrum Analyzer as Illustrated in Figure 8-2

Frequency Modulation

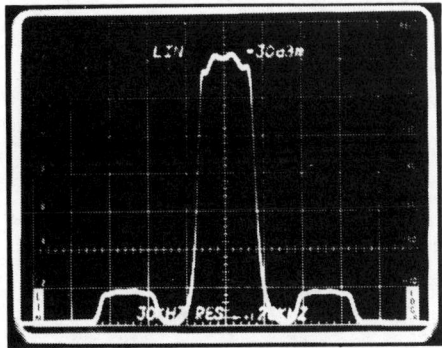

Figure 8-4 Narrowband FM in Frequency Domain

Figure 8-4 is a frequency-domain representation of a narrowband FM signal. The carrier frequency of 275 MHz is determined from the spectrum analyzer center-frequency dial. The modulating frequency is equal to the frequency spacing between the carrier and either of the first sidebands. At a span of 20 kHz/DIV and a sideband spacing of 2.5 DIV, the modulating frequency is 50 kHz. To determine the deviation ΔF, utilize the fact that the amplitude of the carrier is represented by the magnitude of the Bessel function of the first kind, zero order, with modulation index equal to argument; while the amplitude of either of the first sidebands is represented by the magnitude of the Bessel function of the first kind, first order and modulation index equal to the argument. Mathematically,

$$A_c \propto J_0 \left(\frac{\Delta F}{f} \right)$$

$$A_{s1} \propto J_1 \left(\frac{\Delta F}{f} \right)$$

In order to make an actual calculation, an equality rather than a proportional relationship is needed. This is obtained by taking the ratio between the two amplitudes; thus,

$$\frac{A_{s1}}{A_c} = \frac{J_1 \left(\frac{\Delta F}{f} \right)}{J_0 \left(\frac{\Delta F}{f} \right)} \tag{8-1}$$

where the ratio is of voltages rather than powers. The easiest way to determine the deviation is now to use the fact that for small modulation indices the following relationship holds:

$$\frac{J_1\left(\frac{\Delta F}{f}\right)}{J_0\left(\frac{\Delta F}{f}\right)} = \frac{\Delta F}{2f} \tag{8-2}$$

A variation of this is given in Equation 4-25. From Figure 8-4, the voltage ratio of the amplitudes is $0.9/7.2 = 0.125$. Hence,

$$\frac{J_1\left(\frac{\Delta F}{f}\right)}{J_0\left(\frac{\Delta F}{f}\right)} = \frac{\Delta F}{2f} = 0.125$$

and the modulation index is $\Delta F/f = 0.25$.

Since the modulation frequency $f = 50$ kHz, the deviation $\Delta F = 0.25(50) = 12.5$ kHz. If one does not recall the formula, or if it is desired to try for a more accurate result because the formula becomes inaccurate above modulation indices of about 0.5, the procedure is to use tables of Bessel functions. Here, the first step is to let the $J_0(\Delta F/f)$ term equal unity, an obviously reasonable assumption for small values of argument, as demonstrated by the graph of Figure 8-1. The second step is to find in the table the value of $J_1(\Delta F/f)$, which is equal to the measured ratio of sideband to carrier. With this as a first approximation, the calculation can be refined by checking for a closer match of the $(\Delta F/f)$ value in the vicinity of the first approximation. For example, suppose the voltage ratio of first sideband amplitude to carrier amplitude, as determined from the spectrum analyzer display, is 0.375. *Assuming that $J_0(\Delta F/f) \cong 1$, $\Delta F/f \cong 0.8$, as is illustrated in the partial table of Bessel functions, Figure 8-5. Checking actual ratios of

$$\frac{J_1\left(\frac{\Delta F}{f}\right)}{J_0\left(\frac{\Delta F}{f}\right)}$$

in the vicinity of $\Delta F/f = 0.8$, note that at $\Delta F/f = 0.7$,

$$\frac{J_1(0.7)}{J_0(0.7)} = \frac{.329}{.881} \cong 0.373$$

The modulation index is, therefore, much closer to 0.7 than to 0.8.

---

*These numbers were chosen to illustrate the method. A ratio measurement to an accuracy of three significant figures is beyond the accuracy of most spectrum analyzers.

Frequency Modulation

Table 8-1
Partial Table of Bessel Functions

$\dfrac{\Delta F}{f}$	0.6	0.7	0.8	0.9
$J_0\left(\dfrac{\Delta F}{f}\right)$	0.912	0.881	0.846	0.808
$J_1\left(\dfrac{\Delta F}{f}\right)$	0.287	0.329	0.369	0.406
				↑0.375

8.3. WIDEBAND FM

Most FM measurements are for modulation indices of about 1 to 10, which for purposes of this discussion has been designated wideband. As for small modulation indices, the method of measurement depends on the determination of ratios. This creates difficulties because in many instances it is important to determine the deviation to a very high degree of accuracy, whereas the performance of most spectrum analyzers precludes the measurement of relative amplitude to better than two significant figures. There is only one relative-amplitude determination that can be obtained to a high degree of accuracy; this is where one of the components is zero. The technique where one of the amplitude components, usually the carrier, is made to go to zero is known as the carrier-null method, the Bessel null method, or the Crosby null method — after Murray G. Crosby, who did much of the basic work on FM measurements.

The carrier null method of FM deviation measurement is the most used and also the most accurate of all methods. However, it is only applicable in cases where the FM signal can be changed during the measurement. Where the signal to be measured is fixed, it is necessary to use other, more complicated and less accurate, measurement techniques.

Most people have no problem with the fact that for a sinusoid, such as $\cos\theta$, there are specific values of the angle θ, where the magnitude of the sinusoid goes to zero. For $\cos\theta$, this occurs when the angle θ is an odd multiple of $\pi/2$, that is, $\theta = 90°, 270°$, etc. A similar relationship occurs for Bessel functions, where the magnitude of the function goes to zero at certain specific values of the modulation index. These points of zero amplitude are called nulls and, when referring specifically to the carrier component, the term carrier null is used. Carrier nulls occur at those modulation indices where the zero-order

Bessel function of the first kind goes through zero. These points of zero carrier amplitude can be seen in Figure 8-1, where the $J_0(t)$ curve crosses the zero axis.

Table 8-2 shows the first 10 modulation indices at which the carrier goes through a null. The actual measurement procedure will now be illustrated by means of examples.

Table 8-2
Table of Carrier Nulls

Carrier Null	Modulation Index ($\Delta F/f$)
First	2.4048
Second	5.5201
Third	8.6537
Fourth	11.7915
Fifth	14.9309
Sixth	18.0711
Seventh	21.2116
Eighth	24.3525
Ninth	27.4935
Tenth	30.6346

8.3.1. Example 1

Figure 8-5 shows a spectrum analyzer display of a wideband FM signal. The sidebands are three-quarters of a graticule division apart and, at 10 kHz/DIV, this corresponds to a modulation frequency f of 7.5 kHz. The object is to determine the deviation. This will be accomplished by changing the modulation frequency f such that the modulation index $\Delta F/f$ corresponds to one of the carrier nulls. The deviation will then be computed from the known values of f and $\Delta F/f$. One of the dangers in this procedure is mistaking one carrier null for another. Thus, one might think that the display corresponds to the first carrier null at a modulation index of 2.4, whereas the actual modulation index is 5.5, corresponding to the second carrier null. The following procedure will guard against such an error.

First, guess at the possible limits of the deviation. From Figure 8-5 it is observed that the sideband amplitudes start falling off at about 30 kHz from the carrier, which is at the center of the display; and there are virtually no sidebands beyond 40 kHz from the carrier. The guess, therefore, is that the modulation index is probably around $\Delta F/f = 30/7.5 = 4$, but it might be as high as $\Delta F/f = 40/7.5 = 5.33$. Both the best and maximum-limit guesses are between the first and second carrier nulls, which are at a modulation index of 2.4 and 5.5 respectively. If correct, the first carrier null should be able to be obtained by increasing the

Frequency Modulation

modulating frequency f while maintaining the modulating signal amplitude constant.* As the modulating frequency is increased, the carrier amplitude is seen to decrease until it goes to zero, as shown in Figure 8-6. In this example, it occurs at a modulating frequency of 10 kHz. Since the first carrier null occurs at a modulation index of 2.4, the deviation is $\Delta F/f = 2.4$, $\Delta F = 2.4 \cdot 10 = 24$ kHz.

Just to make sure that there is no mistake, the modulating frequency could be decreased to check the deviation at the second carrier null. The two results should of course agree; otherwise there is an error in the measurement. This is the most accurate method of determining an unknown deviation, since the modulation frequency can be measured to a very high degree of accuracy with a counter.

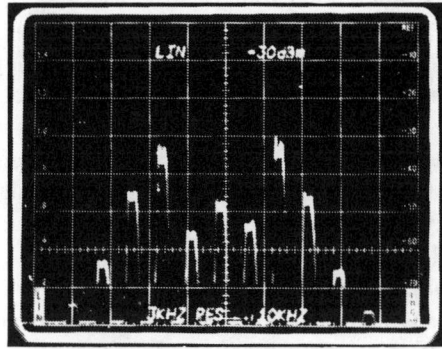

Figure 8-5 FM Deviation Measurement

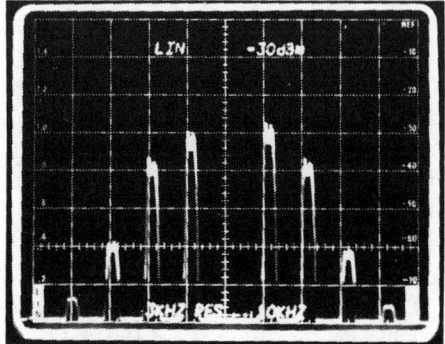

Figure 8-6 Deviation Measurement Illustrating Carrier Null

---

*For a linear modulator, the deviation is dependent on only the amplitude of the modulating signal.

8.3.2. Example 2

Another method of getting a carrier null is to change the amplitude of the modulating signal source. The deviation is directly proportional to the modulating voltage amplitude when operating within the linear range of the modulator. Voltages, however, cannot be measured as accurately as frequencies, so this method is less accurate than that based on frequency measurement.

Frequently, the object is not to determine what the deviation is but, rather, to establish a particular deviation. Such a case might be found in the broadcast industry where the Federal Communication Commission specifies 200 kHz channel separation with a maximum frequency deviation of 75 kHz with respect to the carrier and at a maximum modulating frequency of 15 kHz. It would be of interest to adjust such a transmission system so that the modulating voltage would not exceed the level corresponding to a 75 kHz deviation. The simplest way of establishing the level of the maximum voltage not to be exceeded is to set the modulating frequency such that a deviation of 75 kHz will correspond to a carrier null and then adjust for the null by changing the modulating voltage amplitude. Thus, at a deviation of 75 kHz, the first carrier null corresponds to a modulating frequency of $f = 75/2.4 \cong 31.2$ kHz. This is greater than the maximum permitted modulating frequency of 15 kHz, so the first carrier null cannot be used here. The second carrier null results in a computed modulating frequency of $75/5.5201 = 13.586$ kHz, which can be used. The sequence of photographs in Figure 8-7 illustrates the procedure.

With the modulating frequency set to 13,586 Hz by means of a counter, the modulating voltage is increased to obtain the second carrier null. This corresponds to a deviation of 75 kHz. Figure 8-7(A) shows the unmodulated carrier corresponding to zero modulating voltage. As the modulating voltage is increased, the carrier amplitude decreases and sidebands appear as shown in Figure 8-7(B). As the modulating voltage is increased further, the first carrier null corresponding to a modulation index of 2.4 and a deviation of 32.5 kHz is reached. This is shown in Figure 8-7(C). As the modulating voltage is increased even further, the carrier amplitude increases again and more sidebands appear, as is shown in Figure 8-7(D). Finally, in Figure 8-7(E) with the modulating voltage increased yet again, the second null is reached, corresponding to a frequency deviation of 75 kHz. As long as the final voltage setting is not exceeded, the transmitter will operate within the permitted limit of 75 kHz deviation.

8.3.3. Example 3

Deviation linearity is a measure of the nonlinearity, existing in an FM transmitter or signal generator, between the carrier frequency deviation and the voltage

Frequency Modulation

amplitude of the modulating frequency causing the deviation. It is described graphically as the ratio of the modulating voltage divided by the modulation index to the modulation index, as is shown in Figure 8-8.

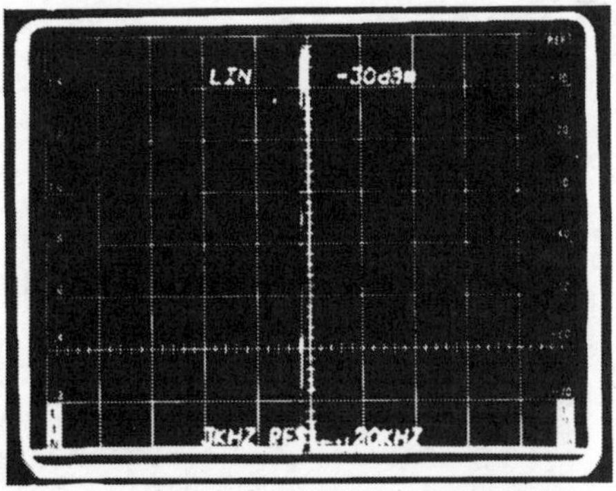

(A) Unmodulated Carrier

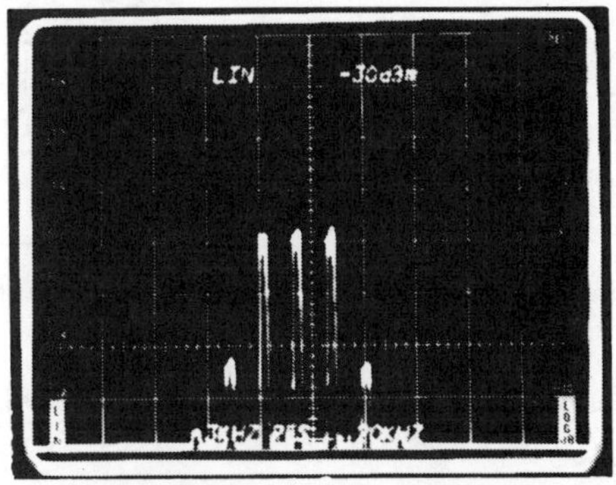

(B) Small Degree of Modulation

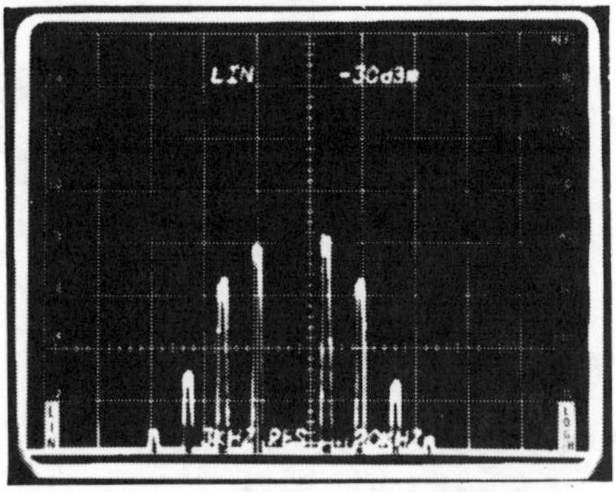

(C) Approaching First Carrier Null

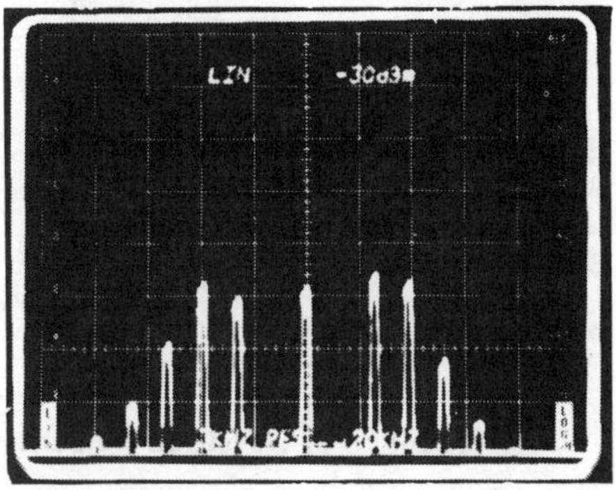

(D) Between First and Second Carrier Null

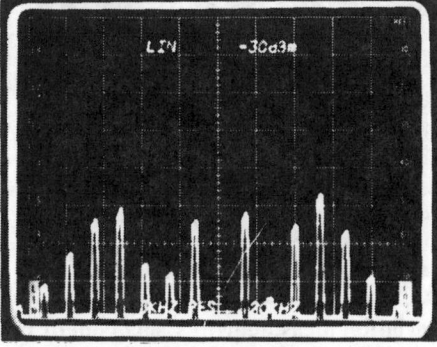

(E) Approaching Second Carrier Null

Figure 8-7 Carrier Null Method of Deviation Adjustment in FM Using Variable Amplitude Modulating Signal

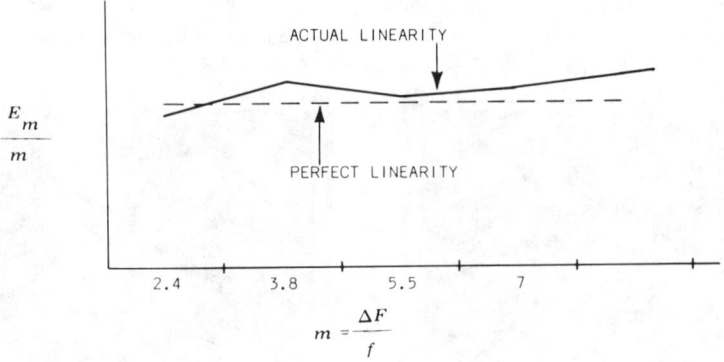

Figure 8-8 Graphic Display of Deviation Linearity

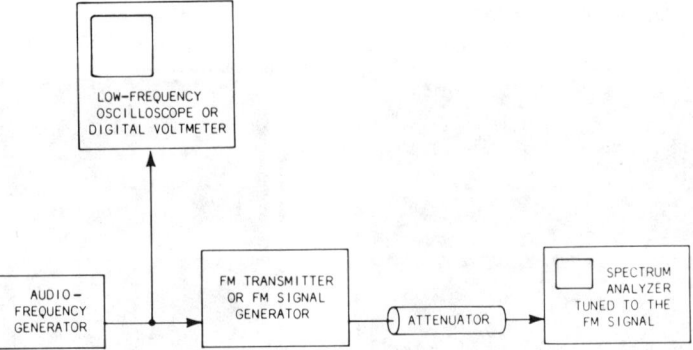

Figure 8-9 FM Deviation Linearity Measurement

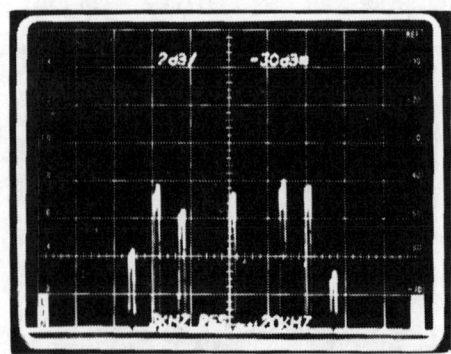

Figure 8-10 FM Deviation Measurement Using Null of First Sideband

Frequency Modulation

Table 8-3
Table of First Sideband Nulls

First-Sideband Null	First	Second	Third	Fourth	Fifth	Sixth	Seventh	Eighth	Ninth
Modulation Index ($\Delta F/f$)	3.83	7.02	10.17	13.32	16.47	19.62	22.76	25.90	29.05

The measurement consists of measuring the voltage amplitude of the modulating frequency for successive carrier and sideband nulls for as many values of modulation index as desired. A spectrum analyzer capable of high resolution and low incidental FM is used to display the signal nulls, while the amplitude of the modulating frequency may be measured accurately with a low-frequency oscilloscope or digital AC voltmeter. See Figure 8-9.

Ratios of modulating-frequency voltage divided by modulation index are plotted vertically and the values of modulation index are plotted horizontally to graphically display the degree of nonlinearity that may be present. Ideally, the curve should represent a horizontal straight line for the complete range of modulation index values.

Figure 8-8 shows a graphical representation of the nonlinearity measured on a typical klystron high-frequency FM oscillator. The amount of nonlinearity is not affected by changing modulation frequencies.

Sometimes it is either inconvenient or impossible to obtain a carrier null. The method, then, is to null one of the sidebands, preferably the first sideband. The procedures are identical to those already described except that the modulation indices associated with the first sideband are used. Table 8-3 is a table of first-sideband nulls, and Figure 8-10 is a spectrum analyzer display showing the first sideband null at a modulation index of 3.83.

8.4. ULTRAWIDEBAND FM

There are special problems in determining the frequency deviation of ultrawideband FM. These are:

1. Inability to resolve the separate frequency components of the signal because the modulation frequency is less than the narrowest spectrum analyzer resolution bandwidth.
2. Even when the sidebands can be resolved, there is still difficulty in identifying the carrier among the many (sometimes hundreds) of displayed sidebands.
3. Even when the carrier is identified, it is almost impossible to count through more than about 10 nulls without a large measure of uncertainty about the accuracy of the count.

The simplest method is to consider that the deviation is one-half of the total occupied signal bandwidth as measured on the spectrum analyzer. This is based on the fact that the sideband energy falls off quite fast outside the frequency deviation. An approximate formula derived by Charest is $B/\Delta F = 2.0 + (4/\beta)$, where B is the 40 dB down bandwidth and β is the modulation index. Ignoring the $4/\beta$ term at modulation indices greater than 50 introduces less than a 5% error in the measurement. A more accurate simple formula for modulation indices less than 50 is $B/\Delta F = 2.5 + (4/\beta)$.

Figure 8-11 shows a spectrum analyzer display of an ultrawideband FM signal. Assuming that the deviation is one-half the total 40-dB-down signal bandwidth, 5 DIV · 50 kHz/DIV = 250 kHz is obtained and the deviation is $\Delta F = 125$ kHz. The modulating frequency is 100 Hz, which was determined from a different measurement as will be discussed later in this chapter. Here the modulation index is 1250, so that ignoring the $4/\beta$ term introduces negligible error. For a smaller modulation index, a correction factor computed from the $4/\beta$ term could be added to the original computation. More accurate calculations can be made by using more elaborate formulas. A detailed discussion on such measurements, including graphs, formulas, and sample calculations, will be found in an article by C. N. Charest.*

A note of caution on the signal bandwidth measurement: The accuracy of the 40-dB-down signal-width measurement depends on the resolution-bandwidth skirt selectivity. The wider the 40-dB-down resolution bandwidth, the greater the error. Unless the signal bandwidth is considerably greater (e.g., 10 times) than the resolution bandwidth, the effect of the resolution skirt selectivity should be considered.

8.5. DETERMINING MODULATION RATE FOR UNRESOLVED SIGNAL

Sometimes the modulation rate of a signal, either AM or FM, is less than the narrowest resolution bandwidth of the spectrum analyzer. This means that the modulating frequency cannot be obtained by the usual means of measuring the frequency difference between resolved adjacent signal components. In order to determine the modulating frequency, it is necessary to operate the spectrum analyzer as a time-domain superheterodyne radio receiver with a CRT indicator. This means that the sweeping oscillator is stopped by tuning to the zero span position. The modulation is now detected and displayed on the CRT. When the signal is AM, detection occurs at the peak of the resolution-amplifier resonance curve. When the signal is FM, detection occurs on the slope of the resolution-amplifier resonance curve as illustrated previously by Figure 8-2. The modula-

---

*Charest, C. N., "Measuring Wide-Bandwidth FM Deviation," *EDN*, March 1, 1969.

Frequency Modulation 143

tion frequency is computed from the measured period of the displayed waveform. Figure 8-12 shows the detected modulating waveform used in the ultra-wideband FM signal of Figure 8-11. Since the period of the waveform is 10 ms, the modulating frequency is 1/10 ms = 100 Hz.

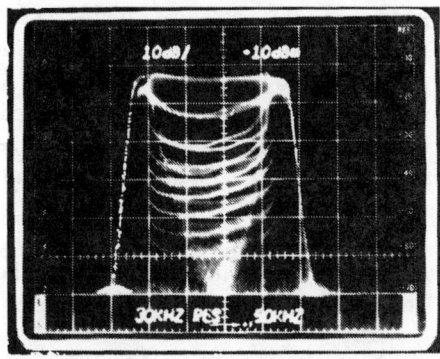

Figure 8-11 Ultra-Wide FM Signal in Frequency Domain

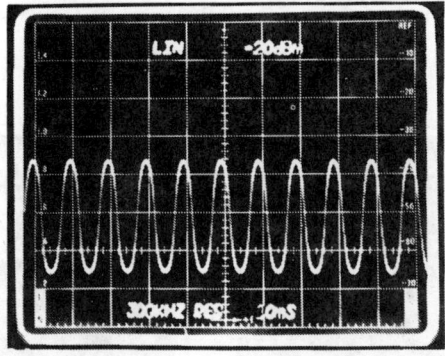

Figure 8-12 Measuring Modulating Frequency at Zero Dispersion

8.6. COMBINED AM AND FM

Combined AM and FM is usually an accidental, or incidental, occurrence. The desired modulation is usually AM, with the FM modulation an incidental by-product of an imperfect AM modulator. Combined AM and FM is characterized by two sidebands of unequal amplitude. This is because the AM sidebands are of the same phase while the FM sidebands are of opposite phase; for a detailed discussion see Chapter 4. Figure 8-13 illustrates the measurement technique for combined AM and FM. Figures 8-13(A) and 8-13(B) show the individual AM and FM spectra which, when generated simultaneously, result in the combined spectrum of Figure 8-13(C). Usually, only the combined spectrum, as shown in Figure 8-13(C), is available. The signal contains both AM and FM because the sideband amplitudes are unequal. Next, since the signal is supposed to be purely AM, we assume that the AM sidebands are larger than the FM sidebands. Except in very unusual circumstances, this will always be the case. A further verification of the small size of the FM sidebands is the fact that the combined signal has only one significant sideband. Now, compute the amplitudes of the individual AM and FM sidebands, using the fact that, in the combined spectrum, one sideband consists of the sum* of an AM and FM sideband while the other sideband consists of the difference between an AM and FM sideband. From Figure 8-13(C), one sideband is about 2.1 DIV high while the other is about 1.7 DIV high. This leads to the conclusion that the AM sidebands are 1.9 DIV high and the FM sidebands are 0.2 DIV high. This is very close to the actual case, as is demonstrated in Figures 8-13(A) and 8-13(B). From the above, now compute:

$$\text{Percentage AM} = \frac{2(1.9)}{6.5} \times 100 = 58.5\%$$

$$\text{FM Modulation Index} = \frac{2(0.2)}{6.5} = 0.0615$$

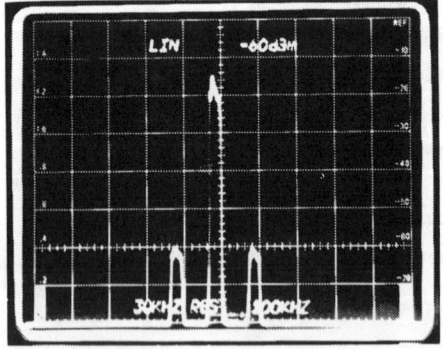 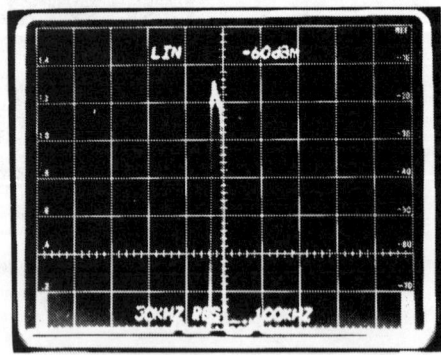

(A) Spectrum Analyzer Display of AM

(B) Spectrum Analyzer Display of Narrowed FM

Figure 8-13 Combined AM and FM in the Frequency Domain

---

*We have made the usual assumption that one FM sideband is fully in phase with the AM sidebands. This is usually the case, but it need not always be so. Therefore, the calculation shows the true AM level, but the FM level could be greater than calculated.

Frequency Modulation

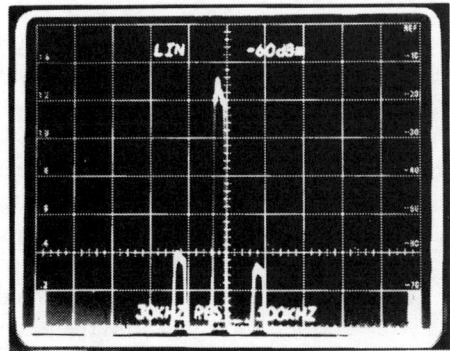

(C) Spectrum Analyzer Display of Combined AM and FM

Figure 8-13 (Continued)

8.7. MULTITONE FM

Unlike AM, where there is no interaction between individual terms, the spectra of multitone FM are complex. These are usually not symmetrical; also, additional sidebands at the sum and difference frequencies of the modulating signals may appear. The frequency-domain appearance of multitone FM is shown in Figure 8-14. Observe that the spectrum is not symmetrical about the carrier component, which is at the center of the screen. Also note that there are additional sidebands besides those due to the 2 kHz and 3 kHz modulating frequencies. A detailed discussion of multitone FM will be found in Giacoletto's "Generalized Theory of Multitone Amplitude and Frequency Modulation."*

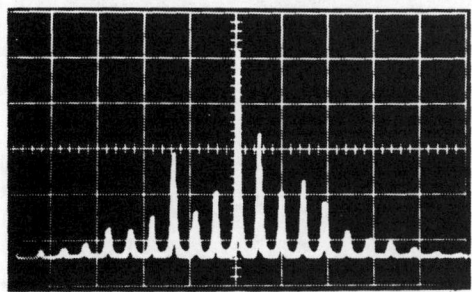

MODULATING FREQUENCIES—2 kHz and 3 kHz
DISPERSION—2 kHz/cm
VERTICAL—LIN

Figure 8-14 Multitone FM in the Frequency Domain

Giacoletto, "Generalized Theory of Multitone Amplitude and Frequency Modulation," *Proc. IRE*, July 1947.

8.8. INTENSIFICATION EFFECTS

Frequently, ultrawideband FM spectra will have an intensified, or brighter, portion in the middle of the spectrum. Such a display is illustrated in Figure 8-15. This intensification effect is due to the performance parameters of the spectrum analyzer and is not indicative of anything about the signal. The effect is generated as the resolution curve of the spectrum analyzer sweeps across the CRT screen. As the FM signal frequency sweeps back and forth, so does the resolution curve of the spectrum analyzer. This back and forth sweeping movement causes two frequency interceptions to occur per FM peak-to-peak deviation. There are, however, two small frequency intervals at the ends of the deviation, where, because of the finite resolution bandwidth, there is only one interception. Consequently, the line density at the ends of the display is half as much as in the middle, causing a difference in brightness. Figure 8-16 is a double exposure of two CW signals showing how the shape of the brightened portion of the spectrum is generated as the FM signal sweeps the resolution curve back and forth.

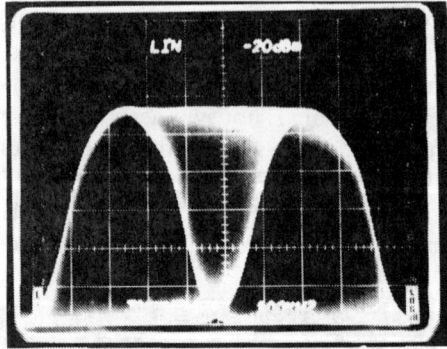

Figure 8-15 Ultra-Wideband FM Showing Intensified Portion in the Center

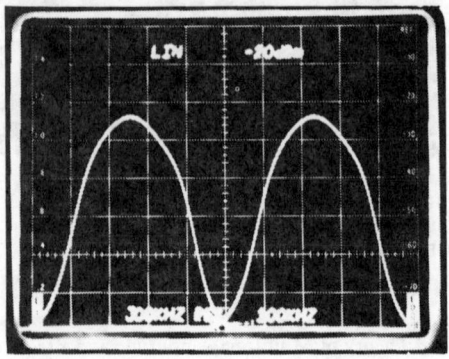

Figure 8-16 As the Resolution Curve Sweeps Back and Forth, the Middle Portion Gets Twice as Many Lines as the Edges

Frequency Modulation

8.9. EXERCISES

8-1. Figure 8-17 shows the spectrum of an FM signal.
   a. What is the modulating frequency?
   b. Is this narrowband or wideband FM?
   c. What is the deviation?

8-2. Figure 8-18 shows the spectrum of a relatively narrowband FM signal with a suppressed carrier. Using the table and graph of Bessel function values found in the appendix, determine the following:
   a. Approximate deviation
   b. Amount of carrier suppression in dB

8-3. Figure 8-19 shows the spectrum of an FM signal. What is the deviation? (Hint: Note the first sideband null.)

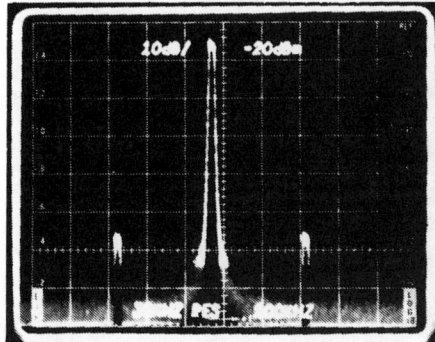

Figure 8-17   Exercise 8-1

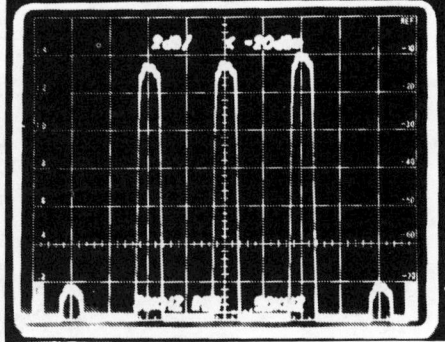

Figure 8-18   Exercise 8-2

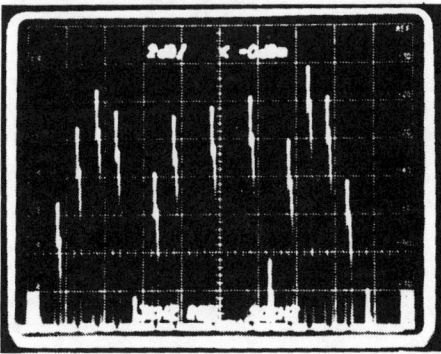

Figure 8-19   Exercise 8-3

Chapter 9
Pulses

9.1. MATHEMATICAL RELATIONSHIPS

There are four possible pulse situations. These consist of dc (noncarrier) pulses or a carrier burst (pulsed RF), in combination with either a discrete Fourier series spectrum due to a pulse train or the dense, Fourier Integral type spectrum characteristic of a single pulse. The sweeping spectrum analyzer can only respond to a pulse train. However, the spectrum analyzer can be set to behave as if dealing with a single pulse. This occurs when the resolution bandwidth is greater than the pulse repetition rate. Thus

$$B \geqslant PRR \tag{9-1}$$

The result is an amplitude spectral density, $S(\omega)$, in units of volts/Hz. Should the instrument bandwidth be less than the pulse repetition rate, the display will consist of Fourier series components of rms amplitude C_n, and will show a discrete line spectrum.

As an example, consider a rectangular pulse of peak amplitude A, pulse width t_o and interpulse interval T. Such a pulse yields the well-known sin x/x spectrum, represented by the generally used relationships shown in Table 9-1.

Table 9-1

	DC Pulse	RF Pulse
Discrete line spectrum (volts rms)	$C_n = \dfrac{2}{\sqrt{2}} \dfrac{At_o}{T} \dfrac{\sin\left(\dfrac{n\pi t_o}{T}\right)}{\left(\dfrac{n\pi t_o}{T}\right)}$	$C_n = \dfrac{At_o}{\sqrt{2}\,T} \dfrac{\sin\left(\dfrac{n\pi t_o}{T}\right)}{\left(\dfrac{n\pi t_o}{T}\right)}$
Dense Continuous Spectrum (volts rms/ Hz of impulse bandwidth, B_i)	$S(\omega) = \dfrac{2At_o B_i}{\sqrt{2}} \dfrac{\sin(\pi f t_o)}{\pi f t_o}$	$S(\omega) = \dfrac{At_o B_i}{\sqrt{2}} \dfrac{\left[\sin \pi t_o(f - f_o)\right]}{\pi t_o(f - f_o)}$

However, the RF burst equation is approximate because the lower frequency sidelobes cannot extend below zero frequency, thus aliasing results. The complete expression for the dense spectrum is

$$S(\omega) = \frac{At_o B_i}{\sqrt{2}} \left\{ \frac{\sin(\pi t_o \Delta f)}{\pi t_o \Delta f} - \frac{\sin[\pi(2f_o + \Delta f)t_o]}{\pi(2f_o + \Delta f)t_o} \right\}$$

Approximation error drops rapidly as the number of carrier cycles within the pulse increases. At 15 cycles, the error is only 1%. Therefore, the additional term can be safely ignored except in situations where quite narrow pulses combine with fairly low frequency carriers. An easy mnemonic is that 100 ns contains ten cycles of a 100 MHz carrier.

This chapter concentrates on the dense, pulsed RF spectrum, which is the primary concern in frequency domain pulse measurements. Discrete spectra of pulsed signals is considered under the heading of Waveform Analysis in Chapter 10.

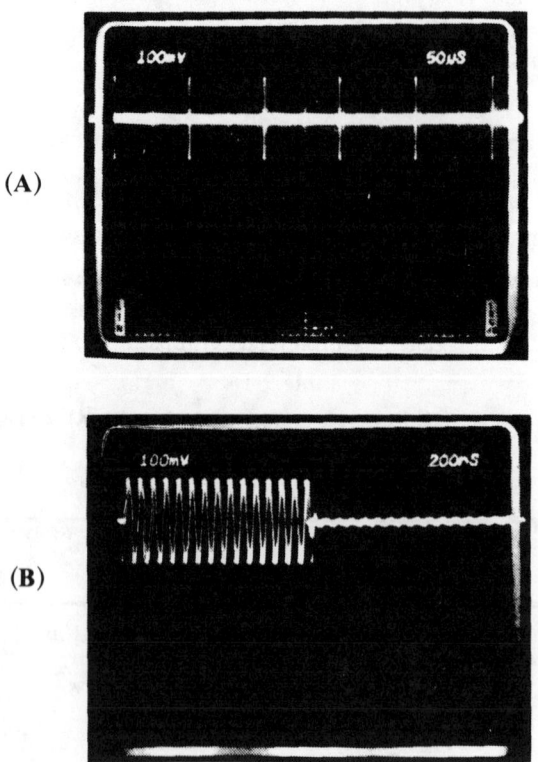

Figure 9-1 Time Domain Appearance of Pulsed RF

Pulses 151

9.2. MAKING THE MEASUREMENT

The basic RF burst is shown in Figures 9-1(A) and 9-1(B) in the time domain and in Figure 9-2 in the frequency domain. The frequency domain relationship between the unmodulated carrier (center screen) and the pulsed RF spectrum is illustrated in Figure 9-3.

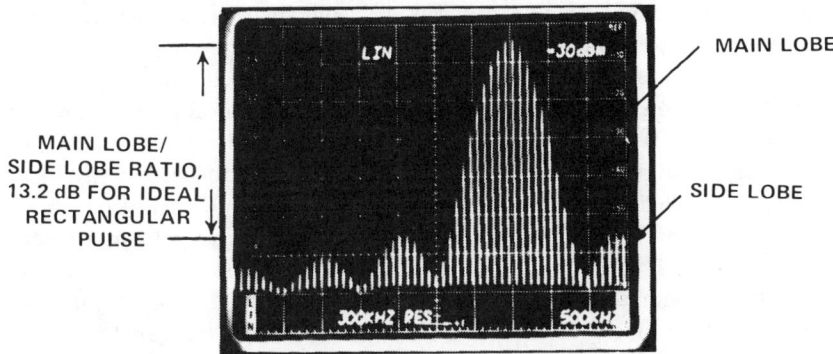

Figure 9-2 Frequency Domain Appearance of Pulsed RF

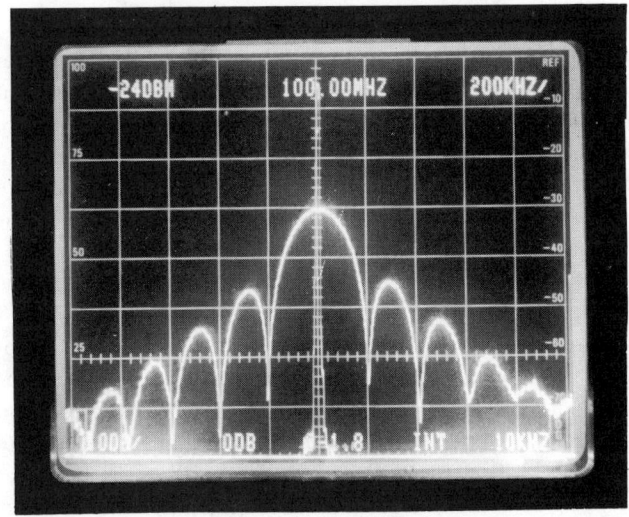

Figure 9-3 Unmodulated Carrier at Center Screen; Note the Loss in Display Amplitude for the Sin x/x of the Pulsed Carrier

The spectrum analyzer will provide a spectral density display when the intercepting bandwidth is greater than the pulse repetition frequency in accordance with Equation (9-1). However, other parameters must be considered in order to determine spectrum shape or amplitude.

1. Set the resolution bandwidth greater than the pulse rate $B \geqslant PRR$ for a dense continuous spectrum. The display is an amplitude spectral density in units of volts/Hz of impulse bandwidth. The spectrum analyzer scale is calibrated in rms units, hence the result is in volts rms. This is true even though current spectrum analyzers invariably use peak detectors. The calibration factor is simply a division by $\sqrt{2}$ on the assumption that we are dealing with sinusoids.
2. The display consists of vertical lines. Each line is an intercept of an incoming pulse, as is shown in the time-frequency diagram (Figure 5-7). The envelope of these so-called repetition rate lines is the desired spectral density of the incoming signal. It has been found that it takes at least five pulse intercepts, or rep-rate lines, per minor lobe and 10 for the main lobe to adequately define the spectrum shape. For a spectral display consisting of one major lobe and two minor lobes, this means 20 input pulses per spectrum analyzer sweep. Hence, for proper definition of the spectrum shape, it is necessary that $20/PRR \leqslant 10$ (time/DIV). The display in Figure 9-2 has about twice the minimally acceptable rep-rate line density.

   As previously noted, these lines are not Fourier series components described by the equations for C_n. Each of these lines is the result of the intercepting of a single input pulse, and the display is of the spectrum analyzer amplifier's transient response. Line spacing does not depend on the frequency span setting of the instrument but rather on the sweep time, or time/DIV. The individual lines are not of interest except when working on pulse-rate-related aspects of the spectrum. The primary interest is in the envelope formed by the peaks of these lines. With the advent of digital storage displays, the internal rep-rate line detail is frequently eliminated, resulting in the smooth envelope display shown in Figure 9-3.
3. The equation for the spectral density, $S(\omega)$, is based on the assumption of an infinitesimally narrow measuring bandwidth. Indeed, for perfect spectrum shape reproduction, it is necessary to convolve with an impulse, as is discussed in Chapter 5. A generally accepted rule is that for accurate spectrum shape definition, it is necessary that the pulse width, t_o, impulse bandwidth, B_i, product be less than one-third. Not much is gained for a product less than one-tenth. Therefore, ideally $0.1 \leqslant t_o B_i \leqslant 0.3$. Using equations developed by Metcalf,* it can be shown that the amplitude error due to the assumption that the resolution bandwidth is an impulse, is only 0.01 dB at $t_o B_i = 0.1$,

---

*Metcalf, "Investigation of Spectrum Signature Instrumentation," *IEEE Trans. EMC-7,* no. 2, June 1965.

Pulses

0.2 dB at $t_o B_i = 0.3$, and 2.1 dB at $t_o B_i = 1.0$. In addition, the spectrum shape becomes noticeably distorted as $t_o B_i$ goes above about 0.5.

4. The main lobe peak for a rectangular shaped carrier burst has an amplitude spectral density of $S(\omega) = A t_o B_i/\sqrt{2}$ (assuming more than 15 carrier cycles within the pulse). The rms unmodulated carrier amplitude is $A/\sqrt{2}$. The ratio of these is known as the pulse desensitization factor and designated by alpha. Thus $\alpha = t_o B_i$ and

$$\alpha \text{ dB} = 20 \log t_o B_i \qquad (9\text{-}3)$$

This equation is quite accurate up to alpha levels of 0.3 and fairly accurate up to 0.5.

It has been shown (see Sabaroff*) that for a gaussian filter, the impulse bandwidth, B_i, is related to the 3 dB and 6 dB bandwidths, B_3 and B_6, by $B_i = 1.06 \, B_6 = 1.5 \, B_3$. It is commonly stated that $\alpha = 1.5 \, t_o B$, where B is the 3 dB bandwidth. This is based on certain assumptions about the spectrum analyzer resolution filter. However, a more exact expression is $\alpha = t_o B_i$. The user is then free to substitute a value or expression for B_i, including $B_i = 1.5 \, B_3$ for certain filters.

The above relationship is illustrated in Figure 9-3. Here we have the unmodulated carrier in the center of the screen, and the pulsed carrier is captured in the second memory of the spectrum analyzer digital storage display. The pulse width is the inverse of the side lobe width or 5 μs, and the impulse bandwidth is approximately equal to the 6 dB 10 kHz bandwidth: α dB = $20 \log 5 \times 10^{-6} \times 10^4 = -26$ dB. This is in essential agreement with the displayed difference of 25 dB. High-accuracy measurements call for better knowledge of the impulse bandwidth. This is discussed later in this chapter.

5. Theoretically the same basic information can be derived from either a time domain oscilloscope display or a frequency domain spectrum analyzer display. For example:

From Figure 9-1(A), observe that the period of the pulse train is $2 \times 50 = 100$ μs. From Figure 9-1(B), observe that the pulse width is $5 \times 0.2 = 1$ μs, and the carrier frequency is $3/0.2$ μs = 15 MHz. The pulse shape is rectangular.

From Figure 9-2, note that side lobe width is $2 \times 0.5 = 1$ MHz, yielding a pulse width of $1/1$ MHz = 1 μs. At five pulse rep-rate lines per division and 500 μs/DIV sweep time (not shown on the CRT readout), the interpulse interval is 100 μs. Likewise, the center frequency, which is read off of a separate front panel dial, shows 15 MHz. Finally, we observe a main lobe to side lobe ratio in linear vertical mode of $6.6/1.5$ or $20 \log 4.4 = 13$ dB, the norm for a rectangular pulse.

---

*Sabaroff, "Impulse Spectrum Analysis for Calibration of Impulse Noise Generation," 1st Conference on RF Reduction – 1954, Illinois Institute of Technology.

Based on the above example, it is apparent that the same basic information can be obtained from both time domain oscilloscope measurements and frequency domain spectrum analyzer measurements. However, spectrum analyzers, except in special cases, predominate in this area because the oscilloscope is not able to display the signals in question. Oscilloscopes do not have the frequency range (pulsed RF signals in the gigahertz region are quite common) or the sensitivity (many signals are in the picowatt region) that is needed. Another difficulty with oscilloscope measurements is that frequently the desired information is the occupied frequency width or spectrum shape rather than the time domain pulse width or pulse shape. While it is theoretically possible to convert from one to the other by means of Fourier mathematics, the task can be quite difficult.

9.3. RADAR PERFORMANCE

Most pulsed RF measurements occur in radar systems. These involve both the determination of what a radar set is transmitting and the adjustment or calibration of the radar set so that it gives the required output. Spectrum analyzers are also frequently used in testing components such as pulsed magnetrons. The data of interest usually involves the following:

1. Carrier frequency (F),
2. Pulse width (t_0),
3. Pulse repetition rate (PRR), interpulse interval (T),
4. Pulse shape,
5. Occupied signal bandwidth,
6. Percentage missing pulses,
7. Carrier on/off ratio,
8. Presence of FM.

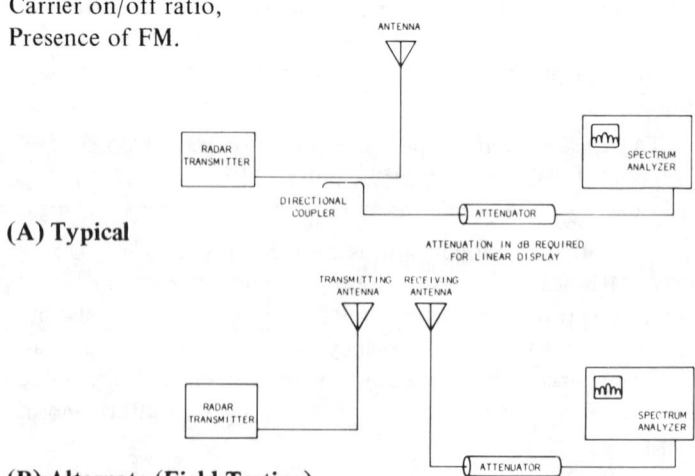

Figure 9-4 Methods of Measuring Characteristics of a Pulsed RF Spectrum by a Radar Transmitter

Pulses 155

Since radar sets usually put out much more power than the spectrum analyzer can accommodate, the signal connection is typically made through a directional coupler. Even then it may be necessary to add attenuation to the signal path to maintain spectrum analyzer operation within the linear region. A typical test setup is shown in Figure 9-4(A). An alternate procedure, especially helpful in radar set tests, as opposed to alignment, is to pick the transmitted signal off the air by a second antenna. Such an arrangement calls for a field transportable spectrum analyzer such as the portable Tektronix Type 492. Care should be exercised in antenna placement to prevent reflections from nearby objects. Figure 9-4(B) shows this alternate test arrangement.

9.4. EFFECT OF PULSE SHAPE

One of the more difficult characterizations of pulsed RF in the frequency domain is that involving pulse shape. While it is theoretically possible to compute a time domain shape from the frequency domain spectrum, it is much easier to match the unknown spectrum against a previously established standard. The graphs of theoretical spectra as a function of pulse shape appearing at the end of Chapter 3 are quite useful for this purpose. While these graphs are based on theory, they are in close agreement with actual observations on a spectrum analyzer. This is illustrated by Figures 9-5 to 9-7.

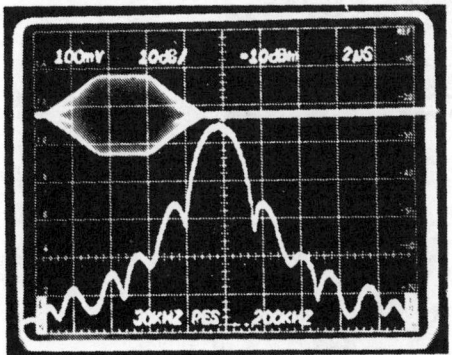

(A) Upper Trace Time Domain Pulse

(B) Lower Trace Frequency Domain Pulse

Figure 9-5 Analyzing Trapezoidal Pulse

Figure 9-5(A) shows the time domain shape of a trapezoidal pulse. It could also be described as a rectangular pulse having appreciable rise and fall time. Figure 9-5(B) is the spectrum of this pulse. The ratio of mainlobe to first sidelobe in Figure 9-5(B) is 20 dB. This is considerably more than the 13-14 dB expected for a rectangular pulse shape but is short of the 26 dB expected for a triangular pulse. Thus, the pulse shape is somewhere between these two extremes. The choice of the precise shape, such as trapezoidal versus cosine-squared, cannot be made on the basis of Figure 9-5(B). For one thing, the spectrum analyzer does not indicate the initial phase characteristics of the signal it displays. Fortunately, the user usually has an idea of what to expect.

Once the user decides that he is dealing with a symmetrical trapezoid, the numbers are easy to compute. From Figure 9-5(B), the average pulse width (t_0) is the inverse of one-half the mainlobe width. Thus, $t_0 = 1/0.2 \cong 5$ μs. From Figure 9-5(A), the average width is $3 + 4/2 = 5$ μs, in good agreement with the computation.

Figure 9-6 shows the time and frequency domain characteristics of a triangular pulse. The mainlobe to sidelobe ratio is about 26 dB. A sidelobe width of 300 kHz corresponds to the roughly 3 μs average pulse width. These numbers are in good agreement with theory.

(A)

(B)

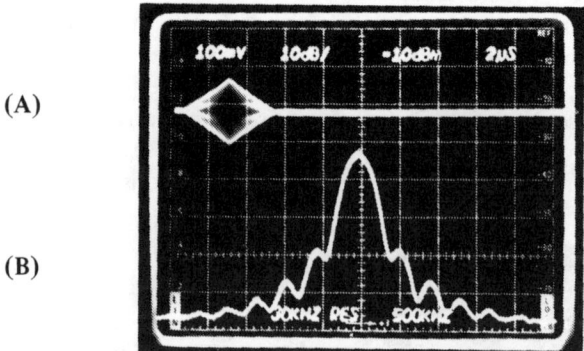

Figure 9-6 Analyzing Triangular Pulse

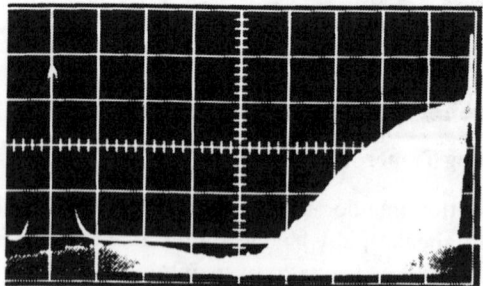

Time Domain: 0.2 μs/DIV
Frequency Domain: 1 MHz/DIV
Vertical: Log

Figure 9-7 Analysis of Sine-Squared Pulse

Pulses 157

A somewhat different pulse-shape analysis problem is exemplified in Figure 9-7. Figure 9-7 is a double-exposure photograph showing the time- and frequency-domain characteristics of a sine-squared pulse. Such pulses are frequently used in the testing of television systems. One of the problems that arises is to ascertain how closely the pulse comes to being perfectly sine-squared. This is difficult to determine from time-domain testing. Small variations in pulse shape are, however, fairly obvious in the frequency domain, since these can have substantial effects on the spectrum. A graph of the theoretical spectrum of a sine-squared, or cosine-squared, pulse is shown in Table 3-2. Figure 9-7 is in good agreement with the theoretically derived spectrum.

9.5. EFFECT OF FM

Pulsed RF signals frequently contain a substantial amount of frequency modulation. The FM can either be intentional, such as in pulse compression radar, or, more usually, an unintentional by-product of pulsing a magnetron or klystron. In any event, the resulting spectra are different from those for non-FM'ing pulsed RF signals. There are three major effects by which one can recognize the spectra of pulsed RF of an FM signal from the spectra of pulsed RF without FM:

(A) Spectra of pulsed RF signals without FM are always symmetrical, have distinct nulls, and the minor lobes are always smaller than the major lobe.
(B) Spectra due to pulsed RF of an FM'ing signal do not have distinct nulls; the sidelobes are larger than for non-FM'ing signals; for nonsymmetrical pulse shapes, such as a sawtooth, the spectrum is usually unsymmetrical.

The effect of FM on pulsed RF was first described in *Radiation Laboratory Series.*[*] Later experiments indicate that the amount of sidelobe lift-up is dependent on the product of pulse width and FM deviation. The greater this product the more the sidelobe lift-up.

The following figures illustrate the frequency-domain appearance of pulsed RF with FM. Figure 9-8 is the spectrum of an FM'ing carrier having 6 MHz peak-to-peak FM deviation, which is pulsed on in 5 μs intervals. Note that the inverse of the pulse width still determines the sidelobe width. Namely, $1/(5~\mu s) = 0.2$ MHz. Similarly, carrier frequency and pulse repetition rate are determined the same way as for pulsed RF without FM. However, the ratio of mainlobe-to-sidelobe size is much less, and the nulls between lobes no longer go down to zero. The pulse width peak-to-peak FM-deviation product is $t_0 \Delta f = 5~\mu s \cdot 6$ MHz $= 30$.

Figure 9-9 is a more extreme example of FM'ing pulsed RF. Here the sidelobes are actually greater than the mainlobe. At a 5 μs pulse width and 10 MHz peak-to-peak FM deviation, the pulse width peak-to-peak deviation product is 50.

---
[*]Montgomery, "Techniques of Microwave Measurements," *Radiation Laboratory Series,* McGraw-Hill or Boston Technical Publishing, Vol. XI.

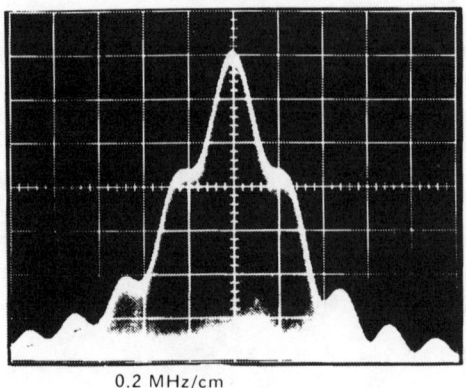

0.2 MHz/cm
VERTICAL — LIN
5 µs PULSE
6 MHz PEAK-TO-PEAK FM

Figure 9-8   Spectrum of Rectangular Pulsed RF with FM

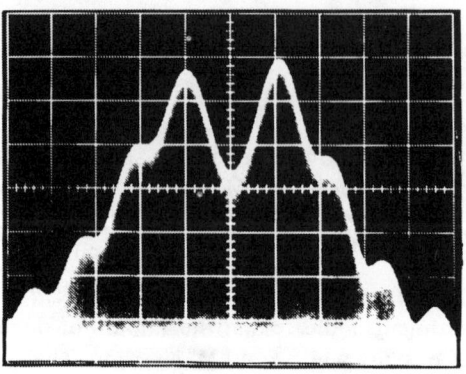

0.2 MHz/cm
VERTICAL — LIN
5 µs PULSE
10 MHz PEAK-TO-PEAK FM

Figure 9-9   Spectrum of Rectangular Pulsed RF with More Severe FM

A plot of the data from Figures 6-8 and 9-9, and data from experiments by Engelson and Breaker,* result in a close approximation to a straight-line relationship between sidelobe lift-up and pulse width deviation product. Figure 9-10 is such a graph where mainlobe-to-sidelobe ratio, normalized with respect to this ratio, in the absence of FM, is plotted as a function of the pulse width peak-to-peak FM-deviation product. The straight-line relationship is approximate, as it is based on insufficient data. This is, however, the best fit available

---

*Engelson and Breaker, "Spectrum Analysis of FM'ing Pulses," *Microwave Journal*, June 1969.

at the present time. The purpose of the graph is to permit an estimate of the amount of FM present. The pulse width is determined in the normal manner from the inverse of the sidelobe width. The normalized mainlobe-to-sidelobe ratio is then determined by measuring the actual ratio and comparing it to the theoretical ratio of 4.6 (13.2 dB) for a rectangular pulse. The peak-to-peak FM deviation is then computed from the $t_0 \Delta F$ product read on Figure 9-10. For example, the output of a pulsed RF magnetron, using rectangular modulation pulses, yields a spectrum having a 3.5-to-one mainlobe-to-sidelobe ratio. The pulse width is 2 μs. What is the peak-to-peak FM deviation? The normalized mainlobe-to-sidelobe ratio is $3.5/4.6 = 0.76$. From Figure 9-10, this corresponds to a pulse width total deviation product of about 14, which corresponds to a peak-to-peak FM deviation of 7 MHz for a 2 μs pulse width.

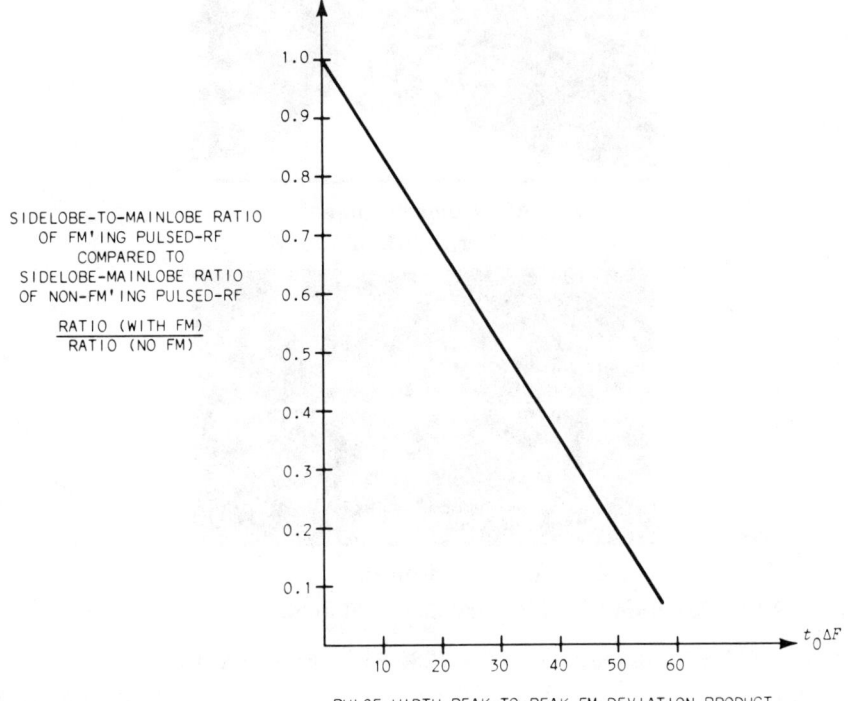

Figure 9-10 Approximate Experimental Relationship Between Sidelobe Mainlobe Ratio and Peak-to-Peak Deviation in FM'ing Pulsed RF

Figure 9-11 shows the effect of FM on pulsed RF having an asymmetrical pulse shape. The spectra of pulsed RF, no matter what the pulse shape may be, is generally symmetrical. However, in the presence of FM, the spectra of pulsed RF of asymmetric pulses are usually asymmetric. Figure 9-11(A) shows the effect of FM on the spectrum of the asymmetrical triangular pulse of Figure 9-11(B). Note that not only have the minima been raised but the spectrum is highly asymmetrical.

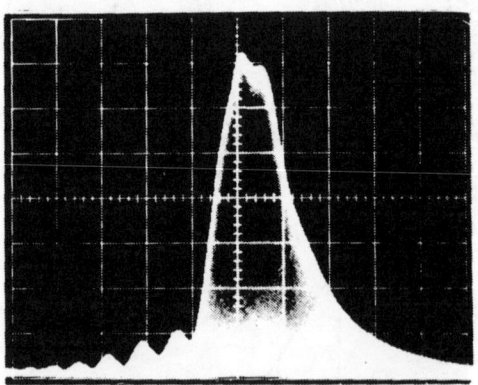

(A) Frequency Domain
Vertical: L in

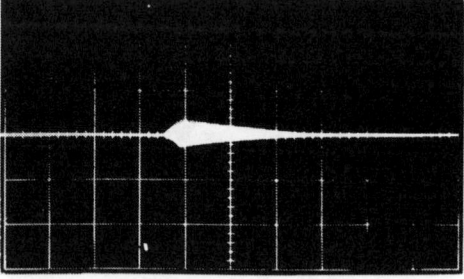

(B) Time Domain

Figure 9-11 Spectrum of FM'ing Asymmetrical Pulse

9.6. PERCENTAGE OF MISSING PULSES DETERMINATION

One of the points of interest in pulsed RF radar measurements is to determine the percentage of misfirings of the oscillator, usually a magnetron. The simplest way of doing this is by observation on a spectrum analyzer, where each of the vertical rep-rate lines corresponds to an oscillator output pulse. The percentage of missing rep-rate lines is, therefore, the same as the percentage of magnetron misfirings. Figure 9-12 simulates the output of a misfiring oscillator. The deliberate misfirings shown in Figure 9-12 are of a repetitive nature while in actual

Pulses 161

situations these would be random; otherwise the two situations are the same. Figure 9-12(A) shows the spectrum of a rectangular-pulse-shape pulsed RF. It will be observed that the signal periodically disappears: these are misfirings. The percentage misfirings is determined by counting the repetition-rate lines. These lines are expanded across the CRT screen by going to a narrow, preferably zero-hertz, span position and a sweep time that permits counting individual lines. This is shown in Figure 9-12(B). Here it will be observed that one-half of the lines are missing. The percentage misfirings, therefore, is 50%. In a real situation, the misfirings would be random. Hence, several photos would have to be taken in order to get a statistically significant number of misfirings.

(A)

(B)

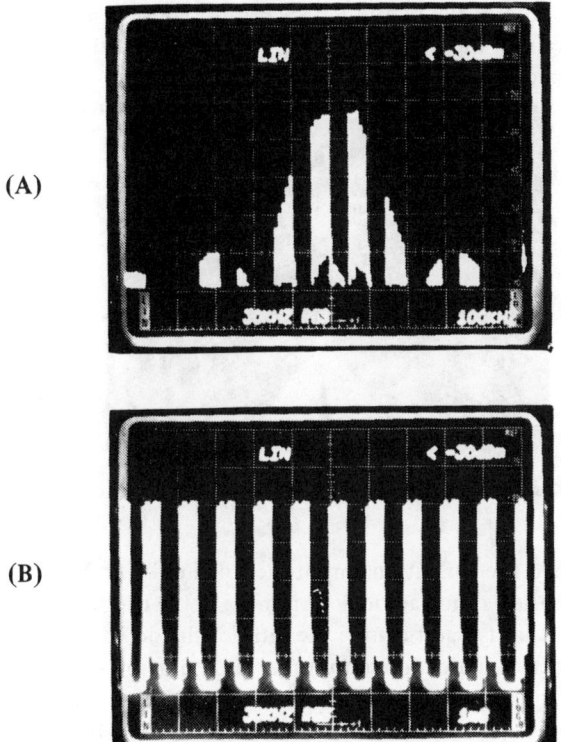

Figure 9-12 Measuring Percentage of Oscillator Misfirings

9.7. MEASURING MODULATOR ON/OFF RATIO

Frequently the modulator or oscillator generating the pulsed RF waveform will not turn off completely, and there continues to be a small amount of output during the interpulse interval. This leads to the need for determining the on/off ratio of the equipment. Figure 9-13(A) shows the time domain appearance of a pulsed signal with a poor on/off ratio. The on amplitude is 4.5 divisions while the off amplitude gives a deflection of 0.9 divisions; hence, the on/off ratio is 4.5/0.9 = 5 or 20 log 5 = 14 dB. To determine the on/off ratio from frequency domain measurements:

1. Figure 9-13(B) is a frequency domain display of the spectrum corresponding to Figure 9-13(A). The spectrum consists of the superposition of two parts; one is basically a typical pulsed RF spectrum while the other part is an ordinary CW spectrum.* We observe that in Figure 9-13(B) the CW part of the spectrum consists of a two-division deflection. The CW response shows up twice: once as a standard resolution curve (the hole in the middle) and again as an addition on top of the pulsed RF spectrum. Subtracting the effect of the CW response from the overall display, it is found that the mainlobe of the pulsed RF spectrum has a two-division deflection.

(A) Time Domain

(B) Frequency Domain

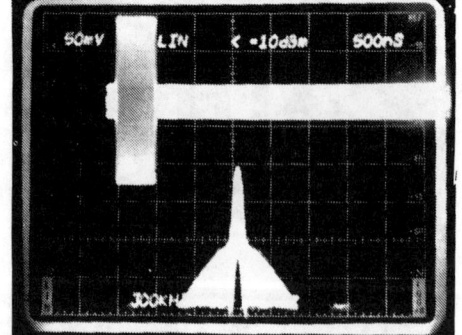

Figure 9-13 Pulsed RF with Poor On/Off Ratio

2. Having determined the deflection amplitudes of the CW and pulsed RF parts of the spectrum, their ratio can be determined; in the example this is 2/2 or 0 dB. Note that what is significant is the ratio of the deflection amplitudes rather than the actual deflection amplitudes. For large on/off ratios, it is necessary to make this measurement in the Log mode.
3. The computation of part two is only part of the answer. This is because the frequency domain display does not show the true relative amplitude difference between the CW and pulsed part of the signal. For pulsed RF, there is a loss in sensitivity relative to CW signals. This loss in sensitivity is $\alpha = t_0 B$.
4. From Figure 9-14, the sidelobe width is 1.7 · 1 MHz = 1.7 MHz. The pulse width is $t_0 = 1/1.7 = 0.6$ μs.

---

*For poor on/off ratios we cannot use the idea of simple superposition. This is because the part of the signal causing the CW spectrum makes the actual pulse height smaller. For on/off ratios greater than about 20 dB, the error due to this effect is negligible.

Pulses 163

5. If the spectrum analyzer in use has calibrated resolution bandwidths, this number is read off, directly, otherwise it has to be measured. For the measurement at hand the resolution bandwidth is calibrated, and is 200 kHz.
6. Now compute the loss in pulse sensitivity: α dB = 20 log(0.6 × 10^{-6}) (3 × 10^5) = −15 dB, where the minus sign denotes a loss.
7. The total on/off ratio is the sum of the two computations; namely, on/off ratio = 15 + 0 = 15 dB. This is in good agreement with the 14 dB computed from time-domain data. Note that when the two numbers are added, two losses were added to find the total loss in sensitivity.

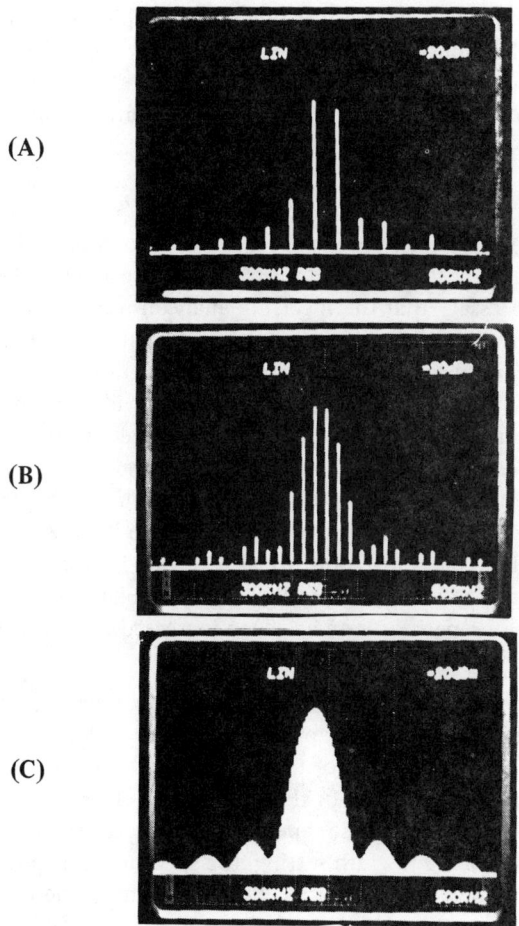

Figure 9-14 Effect of Sweep Time on Spectrum Definition

9.8. EFFECT OF CONTROL SETTINGS

9.8.1. Repetition Rate

Before any measurements can be undertaken, it is necessary to have enough rep-rate samples to define the overall shape of the spectrum. Figure 9-14 illustrates this. Figure 9-14(A) has insufficient rep-rate lines for shape definition. In Figure 9-14(B), the overall shape of the spectrum is just becoming apparent, while the shape in Figure 9-14(C) is very clearly defined. A count of rep-rate lines in Figure 9-14(B) will show less than 5 sample lines per minor lobe and 10 lines for the major lobe. This is usually considered the demarcation line between a defined and undefined spectrum shape. Since the number of lines on the screen is equal to the number of pulses intercepted during one sweep, it is necessary to sweep slower (more time per division) if there are insufficient lines to define the spectrum shape. A smooth pulse spectrum outline will be obtained by use of digital storage when available. Figure 9-3 is such a display.

9.8.2. Dense Versus Line Spectrum

Pulsed RF spectrum analyzer measurements are usually based on a dense-spectrum rather than a line-spectrum interpretation. To achieve a dense-spectrum type of display, it is necessary that the spectrum analyzer resolution setting be greater than the pulse repetition rate. This matter is discussed in considerable detail in Chapter 5. Figure 9-15 illustrates what happens to the appearance of the spectrum as the relationship between resolution bandwidth and pulse repetition rate changes. Figure 9-15(A) shows a standard pulsed RF spectrum. To obtain this display it was necessary that the resolution bandwidth be greater than the pulse repetition frequency. As the pulse repetition rate is increased, it reaches a point where it becomes equal to the resolution bandwidth. This is shown in Figure 9-15(B). This is the transitional spectrum between the dense display of Figure 9-15(A) and the line, or CW, display of Figure 9-15(C). In Figure 9-15(C), the pulse repetition rate has been increased to several times the resolution bandwidth.

The following evolution in the spectrum is observed as the pulse repetition rate is increased. At first, more sample lines appear on the screen, but the spectral shape remains unchanged. The spacing of the lines depends only on the time-per-division setting and is independent of the span setting. This is the normally desired mode of operation. As the pulse repetition rate becomes equal to, and then exceeds, the resolution bandwidth setting, the spectrum and amplitude increases and the lines no longer go down to the baseline. This is the transitional stage. Eventually, new lines going down to the baseline appear. These look like and behave like ordinary CW signals. This shape is easily recognized by the fact that spacing between lines is independent of the sweep time and is determined solely by the frequency span setting.

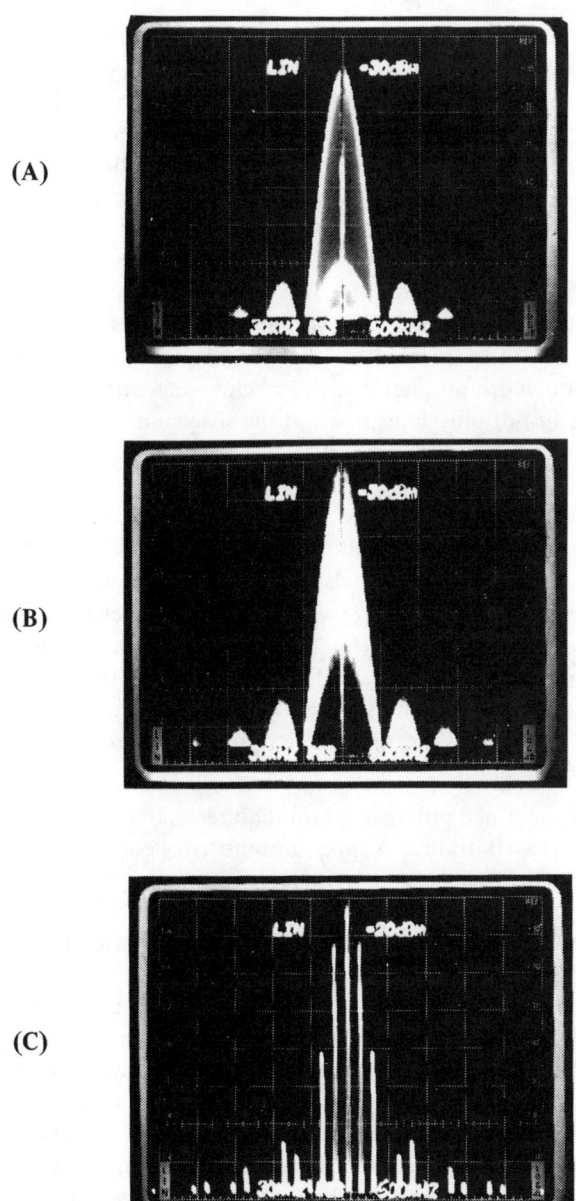

Figure 9-15 Effect of the Pulse Repetition Rate Resolution Bandwidth Relationship on the Type of Spectrum

9.8.3. Fine Detail

In order to display the fine detail of a pulsed RF spectrum, it is necessary that the pulsewidth-resolution-bandwidth product be less than one-tenth. Mathematically: $t_0 B \leq 0.1$. The effect of not meeting this requirement is illustrated in Figure 9-16. In Figure 9-16(A), the pulsewidth-bandwidth product is considerably greater than one-tenth. Note that the sidelobes are almost completely obscured and there are no nulls. In Figure 9-16(B), the pulsewidth-bandwidth product is slightly greater than one-tenth. Here the sidelobes are clearly outlined and the position of the nulls is definite. This permits the unambiguous determination of pulse width. The nulls are, however, not very deep. This makes it difficult to ascertain the degree of incidental FM present. In Figure 9-16(C), where the pulsewidth-bandwidth product is slightly below one-tenth, the nulls are sharp and clear. This not only indicates that the spectrum is properly resolved, but also that no incidental FM is present.

9.8.4. Sensitivity and Dynamic Range

As previously indicated, the sensitivity for pulsed RF signals is less than for CW signals. The ratio between pulsed signals and CW signals of equal peak amplitude (indicated by display level on a linear scale and dB difference on a logarithmic scale) is given by the equation $\alpha = t_0 B_i$. This so-called desensitization factor implies that instrument sensitivity is degraded, but sensitivity in the sense of noise figure remains unchanged. Actually, pulse spreading factor might be a better term because the reduced display amplitude is due to reciprocal spreading, which was discussed in Chapter 3.

In any event, a well-defined pulse spectrum requires that alpha be less than 0.3, representing at least a 10 dB loss in apparent sensitivity compared to a CW signal. Normally, pulsed RF signals are quite large so that amplitude loss is not a problem in itself except that dynamic range is also reduced. The loss in dynamic range occurs because the peak input power for linear operation is little affected by pulse width. About 10 to 15 dB of overdrive is possible for very narrow, submicrosecond pulses as compared to CW signals, but not much more. Consequently, the ability to observe spectra of narrow pulses is determined by the widest resolution bandwidth available and instrument basic sensitivity, or noise figure. For example, a 1 MHz bandwidth combined with a 10 ns pulse represent a loss of $20 \log 10^6 \times 10^{-8} = -40$ dB. An instrument specified at -20 dBm maximum input and -60 dBm sensitivity has an apparent zero dB dynamic range. The spectrum can just barely be displayed due to the greater overdrive for very narrow pulses. A 30 MHz bandwidth yields an alpha of 0.3 and a 10 dB display loss against a -45 dBm sensitivity for a 15 dB improvement in dynamic range. Unfortunately the widest resolution bandwidth available at this time is 3 MHz.

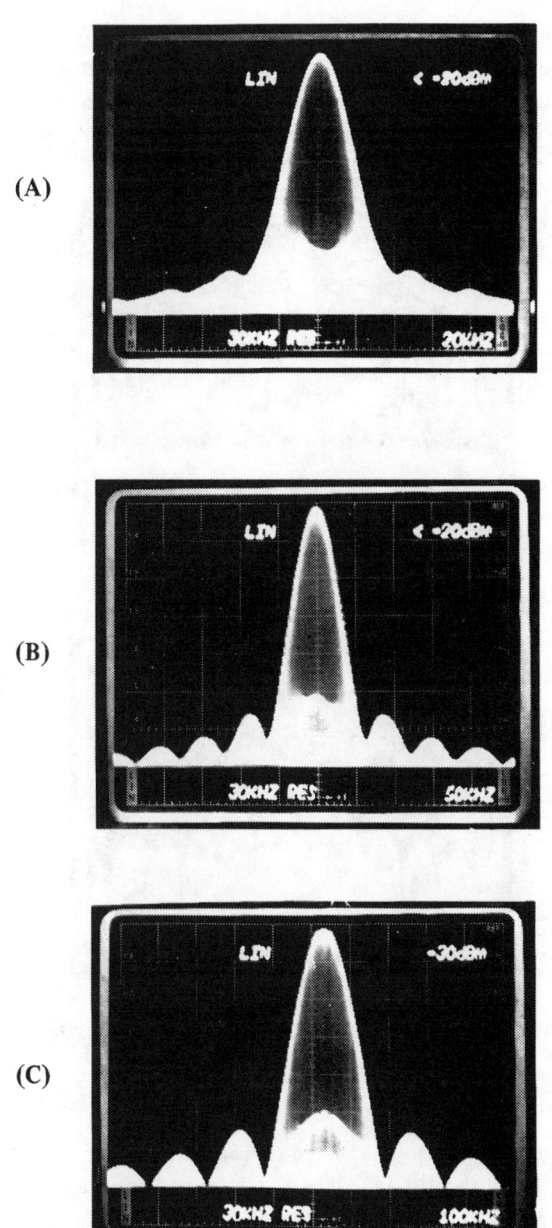

Figure 9-16 Effect of Pulse Width Bandwidth Product on Spectral Display

Figure 9-17 illustrates the change in display level as a function of resolution bandwidth. The spectrum level in 9-17(A) is 20 dB greater than in Figure 9-17(B) because the resolution bandwidth is changed by a factor of 10 from 30 kHz to 300 kHz. The internal noise level in Figure 9-17(A) is 10 dB greater, resulting in a 20 - 10 = 10 dB better dynamic range.

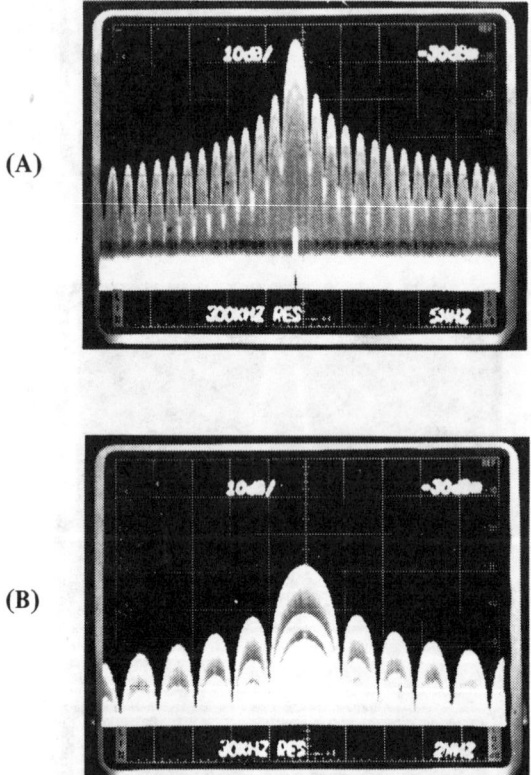

(A)

(B)

Figure 9-17   Illustrating the Loss in Dynamic Range for Narrow Pulses

9.8.5. Display Intensification Effects

Sometimes the spectral display may have extraneous features that are not readily accounted for. Frequently, the peculiarity is in the form of a brightening or intensification of a portion of the spectrum. These intensification effects are almost always due to the operating parameters of the spectrum analyzer. The phenomenon gives no information about the signal and so should be ignored. Two such effects will now be illustrated and the causes explained.

Pulses 169

Figure 9-17(B) is the spectrum of a pulsed RF signal. Note the several bands of intensification. This intensification is caused by the transient response of the variable resolution amplifier.

Figure 9-18 is an expanded version of Figure 9-17(B). We see that each of the rep-rate sample lines, which are due to the variable-resolution-amplifier transient response as explained in Chapter 5, has several stair-like ringing steps. This ringing, which cannot be resolved in Figure 9-17(B), gives the effect of a more intense trace.

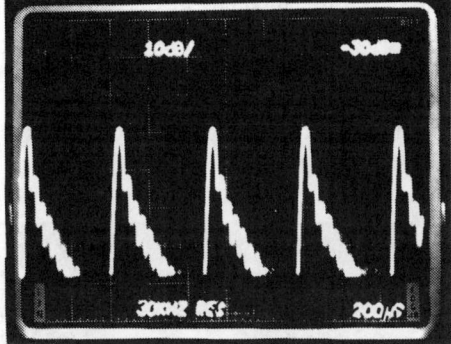

Figure 9-18   Pulsed RF Illustrating Bands of Intensification Across Spectrum

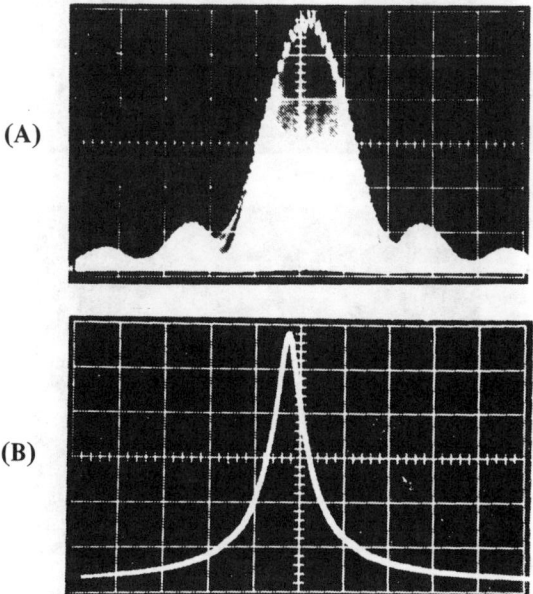

Figure 9-19   Effect of Gradual Resolution Filter Skirt Selectivity on Pulsed RF Spectrum

A different type of intensification effect is illustrated by Figure 9-19(A) and appears to be two spectra. One, quite bright, appears to be a standard (sin x)/x representative of a rectangular pulsed RF. The other, somewhat less intense spectrum, has no sidelobes or minima. This effect is due to gradual rather than abrupt skirts on the variable resolution filter as illustrated in Figure 9-19(B). From Figure 9-19(A), we calculate that the pulse width is roughly 11 μs. In order to have adequate fine detail definition, it is necessary to meet the requirement that $t_0 B < 0.1$. With $t_0 = 11$ μs, the bandwidth has to be less than 9.1 kHz. The resolution curve is quite adequate at the 3 dB or even 6 dB down point, but, at the amplitude level of the spectrum nulls, the resolution curve is much wider. This leads to the combination spectrum of Figure 9-19(A). When the resolution curve shape is more rectangular (Figure 9-20(A)) the nulls are clearly defined, as shown in Figure 9-20(B). Once the cause of the double spectrum in Figure 9-19(A) is understood, the filled-in sidelobes can be ignored and all necessary data obtained.

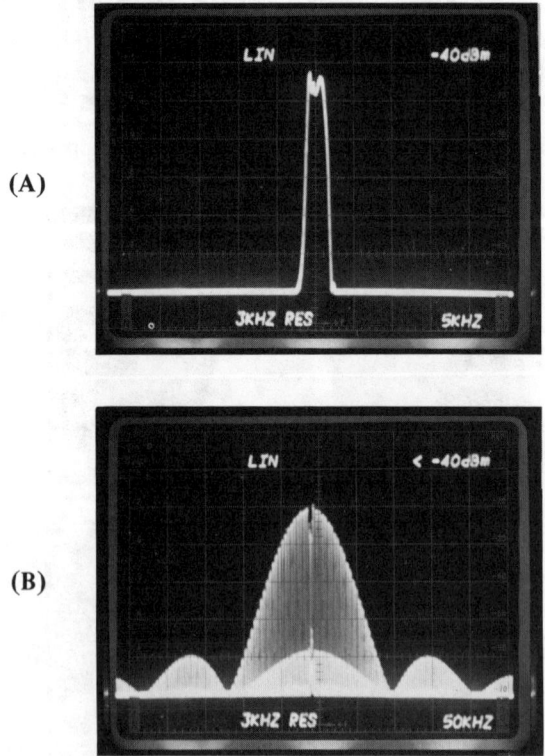

Figure 9-20 Sharp Skirt Selectivity Leads to Well-Defined Nulls

9.9. RECONCILING THEORY AND MEASUREMENT

As indicated in Chapter 3, the use of Fourier transform mathematics without regard to the physical situation can lead to errors. The following example dealing with the frequency distribution of a rectangular pulse will illustrate the relationship between theory and measurement.

The Fourier transform of a rectangular pulse of peak voltage amplitude A and pulse width t_0 is given by Equation (3-27) as

$$F(\omega) = At_0 \frac{\sin \pi f t_0}{\pi f t_0}$$

where $\omega = 2\pi f$. This is the well-known sin X/X spectral distribution.

It is important to recognize that the theoretical spectrum extends over negative as well as positive frequencies. Thus, if one wishes to go back to the original pulse by taking the inverse Fourier transform per Equation (3-26), it is necessary to integrate along the frequency axis from $-\infty$ to $+\infty$. While the negative frequencies are necessary for a generally rigorous mathematical treatment, they have no physical meaning. For a real (no imaginary part) time function — which is the only kind that can be generated by a physical network — it can be shown that the negative frequency spectrum is the conjugate of the positive frequency spectrum. These parts can be added so that the spectrum is considered as consisting of a real part $R(\omega)$ and an imaginary part $X(\omega)$ of only positive frequencies. Thus, for a real time function, Equation (3-26) can be written as:

$$f(t) = \frac{1}{2\pi} \int_{-\infty}^{+\infty} F(\omega) e^{j\omega t} d\omega \quad (9\text{-}4)$$

$$= \frac{1}{2\pi} \int_0^\infty [2R(\omega) \cos \omega t - 2X(\omega) \sin \omega t] d\omega$$

The $R(\omega)$ and $X(\omega)$ terms can be combined by the square root of the sum of the squares procedure similar to that given by Equation (3-2) for the Fourier series. Since $X(\omega) = 0$ for a rectangular pulse, it follows that

$$F(\omega) = 2At_0 \frac{\sin (\pi f t_0)}{\pi f t_0} \quad \text{for } \omega \geq 0 \quad (9\text{-}5)$$

The equation indicates that the peak spectral density at the center of the main lobe for a rectangular pulse signal is $2At_0$ volts/hertz. Multiplying by the resolution bandwidth (B) to eliminate the per unit bandwidth factor and dividing by $\sqrt{2}$ to convert from peak to rms, we conclude that the spectrum analyzer should measure

$$S(\omega) = \sqrt{2} \ At_0 B \frac{\sin (\pi f t_0)}{\pi f t_0} \quad (9\text{-}6)$$

Let us now consider the case of pulsed RF such as illustrated in Figures 9-1 and 9-2. The theoretical expression for the frequency distribution is given by No. 5 of Table 3-2. This expression follows from the application of the frequency shift theorem (No. 8, Table 3-1) to the theoretical distribution of a sinusoid, which is given by Equation (3-11) as $F(\omega) = (1/2)[\delta(f + f_0) + \delta(f - f_0)]$, and the theoretical expression for the frequency distribution of a rectangular pulse. Thus, the Fourier transform of a rectangular pulsed RF signal of pulse width t_0 and peak amplitude A is

$$F(\omega) = \frac{At_0}{2} \left(\frac{\sin \pi t_0(f + f_0)}{\pi t_0(f + f_0)} + \frac{\sin \pi t_0(f - f_0)}{\pi t_0(f - f_0)} \right) \tag{9-7}$$

To determine what the spectrum analyzer will measure, multiply by 2 to eliminate the negative frequency spectrum and divide by $\sqrt{2}$ to convert from peak to rms. The result is

$$S(\omega) = \frac{At_0 B}{\sqrt{2}} \frac{\sin \pi t_0(f - f_0)}{\pi t_0(f - f_0)} \tag{9-8}$$

This is in agreement with Figure 5-9(B) which indicates that the spectrum analyzer indication for a pulsed RF signal compared to a CW signal of equal peak amplitude is $t_0 B$ for an ideal resolution amplifier. The $\sqrt{2}$ term drops out since it appears in the peak to rms conversion of the CW signal also.

An interesting point is to compare the spectral density of the modulated pulse to that of the unmodulated pulse. For this it is necessary to divide the result for the pulsed RF spectrum (Equation (9-9)) by two since the peak-to-peak amplitude of this pulse is 2A. Hence, we are comparing $\sqrt{2} At_0$ with $At_0/2 \sqrt{2}$ which is a ratio of 4.

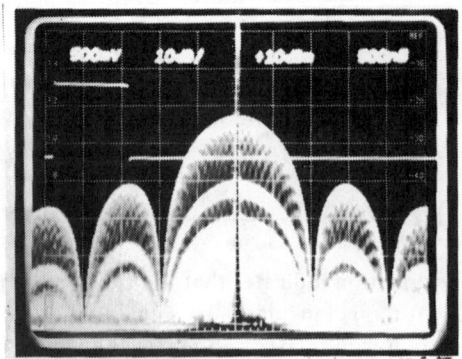

(A) Upper Trace in Time Domain

(B) Lower Trace in Frequency Domain

Figure 9-21 Rectangular Pulse Train

Pulses 173

The above results are illustrated below. Figure 10-21(A) is a rectangular pulse train consisting of 1 μs wide, 1 volt high pulses. The main lobe amplitude should be $\sqrt{2}$ At_0B, which is $\sqrt{2} \cdot 1 \cdot 10^{-6} \times 3 \times 10^4 \approx 42$ mV. Figure 10-21(B) shows that the main lobe height is -14 dBm which is 44.6 mV. Figure 10-22(A) is a pulsed RF signal consisting of 1 μs pulses having a peak-to-peak amplitude of 0.1 volt. We compute a main lobe amplitude of $(0.1 \times 10^{-6} \times 3 \times 10^4)/(2\sqrt{2})$ ≈ 1.1 mV. Figure 10-22(B) shows that the mainlobe height is -46 dBm, which is 1.12 mV. Agreement is well within the experimental accuracy of the equipment.

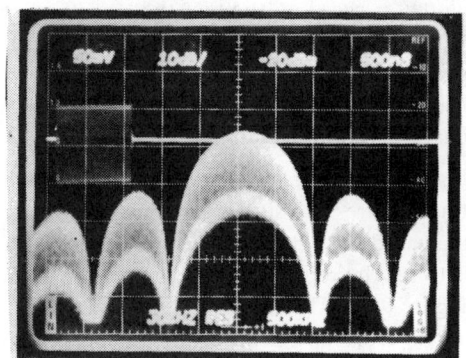

(A) Upper Trace in Time Domain

(B) Lower Trace in Frequency Domain

Figure 9-22   Pulsed RF

9.10.   DETERMINING IMPULSE BANDWIDTH

Knowledge of the impulse bandwidth is an essential prerequisite to the accurate determination of the absolute value of the amplitude spectral density of pulsed signals. The user must determine the impulse bandwidth, since it is normally not specified directly by the manufacturer. The following methods are in sequence of increasing accuracy and also increasing complexity.

1. The simplest technique is to use the resolution bandwidth regardless of how specified. Some manufacturers specify resolution at 3 dB, others at 6 dB. Furthermore, resolution bandwidth is one of the least accurately specified parameters, usually at an uncertainty of ± 20%. This technique therefore is only a crude approximation.
2. The second technique is to use 1.5 times a 3 dB type specification and one times or 1.06 times a 6 dB specification.
3. The third technique is to use whatever multiplication factor the manufacturer recommends for the particular instrument model number.
4. The fourth technique is to measure the actual 3 dB or 6 dB bandwidth of the filter of interest on the actual instrument in use and then apply a multiplication factor (using technique 2 or 3 above). The filter bandwidth is easily determined by displaying a CW signal of any frequency (it could be the calibrator) and checking the frequency width at the appropriate number of dB down.

5. The fifth and most accurate results technique is to actually measure the impulse bandwidth. Many measurement techniques may be found in the literature listed in the reference section at the end of this book. Also the NBS provides an impulse calibration service.

One of the simplest techniques available is also the most accurate. All that is necessary is a source of pulsed RF with good on-off ratio. Results accurate within a few percentage points can be obtained with just moderate care. The technique* is based on the fact that the ratio of center main lobe spectral density, $S_o(\omega)$, to the carrier Fourier line, C_o, for an RF burst is $S_o(\omega)/C_o = B_i f_p$. Display linearity errors can be eliminated by setting $S_o(\omega) = C_o$, and impulse bandwidth, B_i, will simply equal pulse repetition frequency, f_p. The following will illustrate the procedure.

Figure 9-23 is the spectral density of an RF burst using a 10 kHz resolution setting. Figure 9-24 shows the same signal but with the pulse repetition frequency increased so that the display consists of discrete Fourier components. The intent is to set the repetition rate so that the amplitudes are equal. This is illustrated in Figure 9-25. The frequency span has been reduced from 200 kHz/DIV to 2 kHz/DIV so as to spread the display about the center lobe. The light lines are from the dense spectrum stored in one digital storage memory. The CW looking displays are captured in the second digital storage memory. Also, the resolution bandwidth has been reduced by a factor of 10, from 10 kHz to 1 kHz, so that what was a dense spectrum display becomes a discrete display. The pulse repetition rate has been adjusted so that the two types of display are of equal amplitude. Now all that remains is to determine the pulse repetition rate.

A measurement to about 5% can be obtained just from the line spacing across the CRT span. Thus, at 2 kHz/DIV the 4.5 DIV line spacing corresponds to 9 kHz. A more accurate determination requires the use of a frequency counter. This can be an external low-frequency counter that enables the operator to count the modulating signal or a direct counting spectrum analyzer of the newest variety such as the Tektronix 494. The direct count on the spectrum analyzer is illustrated in Figures 9-26 and 9-27. In Figure 9-26, the internal counter has been actuated in the frequency difference mode. The frequency count shows zero. In Figure 9-27, the next Fourier line has been tuned to center screen, and the difference count shows 9.548 kHz.

The whole measurement takes about two minutes once the pulse RF signal is set up. The only precaution to observe is that the display level gain must be the

---

*Engelson, "Check Impulse Bandwidth by Trimming Pulse Rate," *Microwaves,* January 1979.

Pulses 175

same for the two resolution settings used in the measurement. This can be checked by switching between the two settings while displaying a CW signal. An observed change in level can be compensated for in the measurement or the instrument could be recalibrated. Amplitude difference errors between the two spectra can be essentially nil by setting the instrument to a sensitive vertical display position. In the case of this example (Figure 9-25), the setting was 1 dB/DIV, as is shown in the lower left corner.

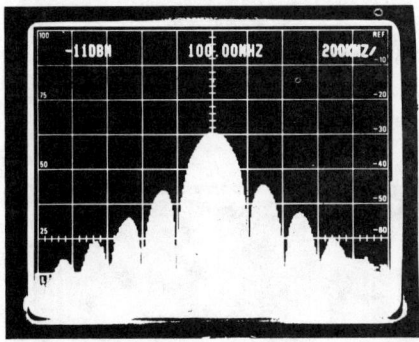

Figure 9-23 Dense Spectrum for Pulsed RF Signal; B > PFF

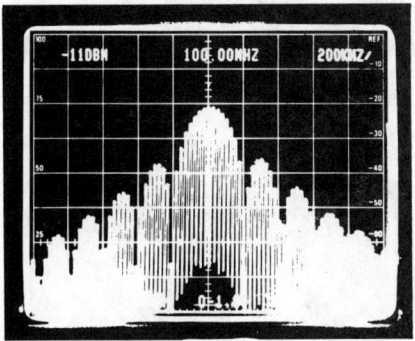

Figure 9-24 Discrete Spectrum for Pulsed RF Signal; B < PFF

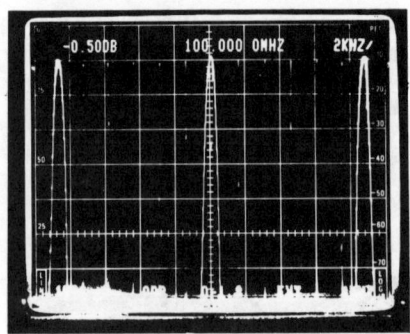

Figure 9-25 Dense and Discrete Spectra Adjusted for Equal Amplitude

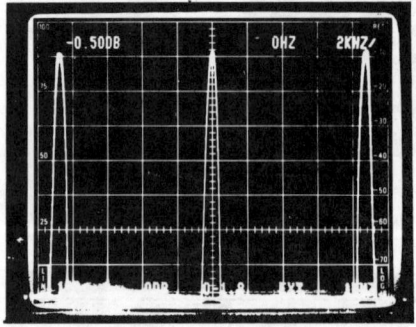

Figure 9-26 The Center Relative Frequency Position Has Been Set to Zero in the ΔF Counter Mode

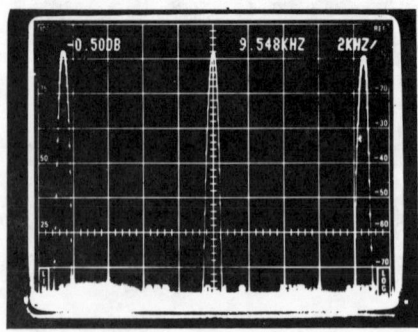

Figure 9-27 The Next Fourier Line Has Been Moved to the Center Showing a Difference Frequency of 9.548 kHz

Pulses

9.11. EXERCISES

9-1. Figure 9-14(C) shows the spectrum of a pulsed RF signal. Determine the following:
   a. Pulse shape
   b. Pulse width
   c. Peak-to-peak pulse height
   d. Widest pulse whose spectrum can be properly displayed without changing spectrum analyzer controls.
   e. Highest and lowest pulse repetition rate compatible with the spectrum analyzer control settings.

9-2. What is the pulse repetition rate for Figure 9-15(C)?

9-3. Compute the on/off ratio for the pulsed RF signal whose spectrum is shown in Figure 9-17(B).

Chapter 10
Miscellaneous Applications

10.1. WAVEFORM ANALYSIS

Fourier series theory, as discussed in Chapter 3, provides a mathematical relationship between the time domain and frequency domain characteristics of various waveforms. The measurement and computational technique is illustrated by the following examples. Note that resolution bandwidth has to be narrower than waveform PRF to resolve the Fourier components.

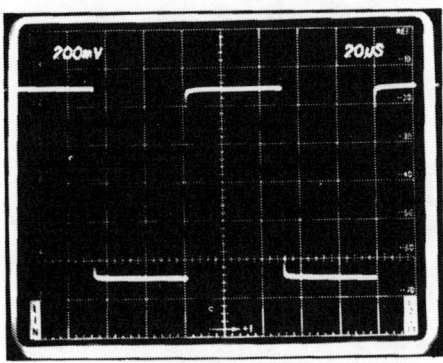

Figure 10-1 Squarewave in Time Domain

10.1.1. Squarewave

Figure 10-1 shows the time domain appearance of a squarewave. The period is 100 μs and the peak-to-peak amplitude is 1 V. According to Fourier theory (see Table 3-2), the squarewave is composed of sinewaves whose amplitudes are given by

$$C_n = \frac{2At_0}{T} \frac{\sin \frac{n\pi t_0}{T}}{\frac{n\pi t_0}{T}} \qquad (10\text{-}1)$$

where C_n is the zero-to-peak ($\sqrt{2}$ rms) amplitude of the nth harmonic of the sinewaves whose superposition makes up the squarewave. The fundamental sinewave frequency (n = 1) has the same period as the squarewave. Hence, $f_0 = 1/T = 1/100 \,\mu s = 10$ kHz. For a symmetrical squarewave, the ratio of pulsewidth to period is $t_0/T = 1/2$. Hence, the amplitude of the fundamental is

$$C_1 = \frac{2A}{2} \frac{\sin \frac{\pi}{2}}{\frac{\pi}{2}}$$

For our example:

$$C_1 = \frac{(1)(2)}{\pi} = 0.64 \text{ V zero to peak or } 0.45 \text{ V rms}$$

$$C_2 = \frac{1}{\pi} \sin \pi = 0$$

$$C_3 = \frac{2}{3\pi} \sin \frac{3\pi}{2} = -0.21 \text{ V peak to peak or } -0.15 \text{ V rms}$$

Also, one can calculate the degree of asymmetry by treating the asymmetric squarewave as a pulse train.*

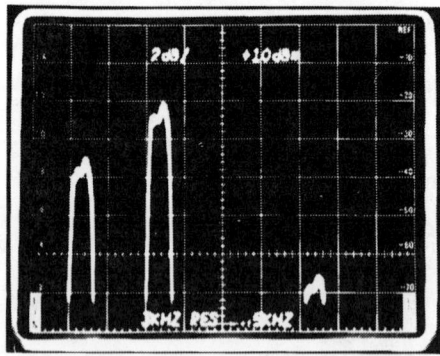

Figure 10-2 Frequency Domain Appearance of Squarewave

The qualitative information on the shape of the time domain waveform can be determined by observing the spectrum on a spectrum analyzer having an appropriate frequency range. Thus, the absence of even harmonics and the fact that the third harmonic is one-third as large as the fundamental, the fifth harmonic

---

*Engelson, "Make the Analyzer Work for You," *Microwaves,* May 1971.

Miscellaneous Applications   181

being one-fifth of the fundamental, etc., indicates that there is a squarewave. The quantitative information dealing with frequencies and amplitudes requires a fully calibrated spectrum analyzer. Figure 10-2 shows the frequency-domain appearance of the squarewave.

The fundamental is two divisions to the right of the zero marker denoting a 10 kHz frequency. Fundamental amplitude is +6 dBm or 0.447 V rms, the third harmonic amplitude is at -3 dBm or 0.158 V rms.

Note that calculated and measured values are in close agreement.

10.1.2. Symmetry Adjustment

A highly useful spectrum analyzer application in the area of waveform analysis is that of symmetry adjustment. Figure 10-3 shows an oscilloscope display of a slightly asymmetrical squarewave. The asymmetry is hardly noticeable. Figure 10-4 shows the same squarewave in the frequency domain. Here, the presence of even harmonics is a clear indication of asymmetry. The signal source could be tuned for best symmetry by adjusting for minimum even-harmonic generation.

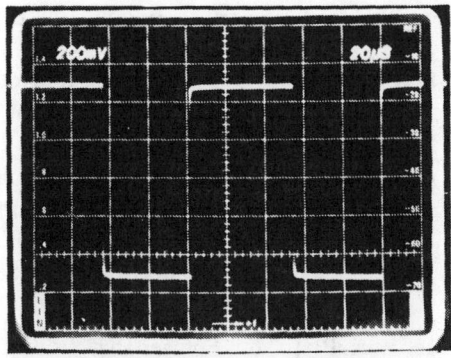

Figure 10-3 Slightly Asymmetric Squarewave in Time Domain

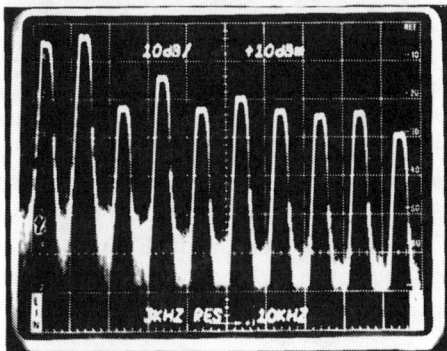

Figure 10-4 Slightly Asymmetrical Squarewave in Frequency Domain (Note Even Harmonics)

10.1.3. Squarewave on Carrier

Consider a sinewave carrier pulsed on and off with a 50% duty cycle as is shown in Figure 10-5. The theoretical spectrum of such a squarewave modulated carrier can be computed in accordance with Fourier series theory in a manner similar to the Fourier integral discussion in Section 9.9. But there is a less rigorous approach.

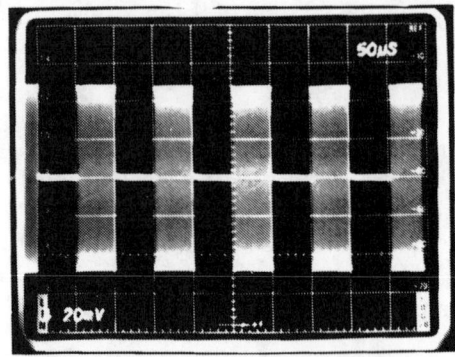

Figure 10-5 Pulsed RF Squarewave in Time Domain

Consider the carrier before it was modulated half the time on and half the time off. Suppose there is a sinewave with a peak-to-peak level of one volt; this is $1/2\sqrt{2}$ V rms. The carrier is squarewave modulated with a duty factor of 1/2. This means that the carrier amplitude after modulation is one-half of that before modulation. The same result follows from a Fourier series analysis, which gives the dc term for a squarewave as 1/2 the peak-to-peak amplitude (see Section 3.2). The dc term is shifted to carrier frequency in accordance with Property 8, Table 3-1.

Next consider the sidebands. Perform a Fourier analysis on the squarewave envelope as in Section 10.1.1. Thus, the peak value of the first sideband is $C_1 = 2/\pi$ and the rms value is $\sqrt{2}/\pi$.

However, the waveform inside the squarewave is sinusoid. So, just as for the case of the continuous sinewave, multiply by the factor $1/2\sqrt{2}$ to get from peak-to-peak to rms. Finally, the spectrum of the squarewave discussed in Section 10.1.1 is one-sided above the dc term. The spectrum of the squarewave modulated sinewave is two-sided, being symmetrically distributed about the carrier. When a single-sided spectrum is split into a double-sided spectrum without a change in energy, the sidebands must be reduced in amplitude. The conservation of energy principle dictates that each of the spectral voltage components be reduced by $1/\sqrt{2}$ so that the energy is reduced by one-half. Thus, each sideband of the

Miscellaneous Applications

straightforward squarewave needs to be multiplied by $1/2\sqrt{2}$ and $1/\sqrt{2}$ to get the amplitudes of the squarewave modulated sinewave spectrum. The multiplication factor of $1/4$ was also derived in Section 9.9 for the Fourier transform situation.

For Figure 10-5, calculate the following rms voltage levels. Carrier amplitude is $(.1/2)(1/2\sqrt{2}) = 17.7$ mV compared to 35.4 mV for an unmodulated sinewave. First sideband amplitude is $(.1/4)(0.45) = 11.2$ mV. Third sideband is $(.1/4)(.15) = 3.8$ mV. Even ordered sidebands are zero.

Figure 10-6 is a spectrum analyzer display of the waveform shown in Figure 10-5. The carrier amplitude is -21 dBm or 19.9 mV; first sideband is -25 dBm or 12.6 mV; and third sideband amplitude is -35 dBm or 4 mV.

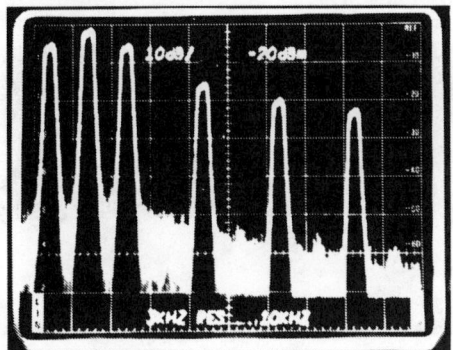

Figure 10-6 Pulsed RF Squarewave in Frequency Domain

10.2. RANDOM NOISE MEASUREMENT

10.2.1. Relative Measurement

The subject of random noise is quite complex. A complete discussion of this subject requires probability mathematics — a subject that is beyond the scope of this volume. Those interested will find many references in this area.* The frequency domain description of random noise is expressed by a power spectrum, where the basic element is spectral density in units of power per unit bandwidth. Besides the need to express noise in units of power per bandwidth, the absolute level indicated by a measuring device is also affected by the detector characteristics.

---

*See, for example, Davenport & Root, *An Introduction to the Theory of Random Signals and Noise,* McGraw-Hill, 1958; also, Blackman & Tukey, *The Measurement of Power Spectra,* Dover Publications, 1958.

The effect on the output of various detector time constants and a summary of useful formulas has been compiled by Peterson.*

However, the above complications do not arise if one is willing to settle for a partial, relative distribution. This expression of the signal is all that is needed for many applications, and it can be easily obtained with any spectrum analyzer.

Figures 10-7 and 10-8 illustrate such a measurement. The illustrated problem required the adjustment of the bias on a zener noise source for the flattest output in the frequency domain. The two oscilloscope presentations, Figures 10-7(A) and 10-8(A), show relative output amplitude, but the frequency distribution cannot be ascertained. The spectrum analyzer displays, Figures 10-7(B) and 10-8(B), clearly show that the bias adjusted for Figure 10-8 results in the flatter output frequency distribution.

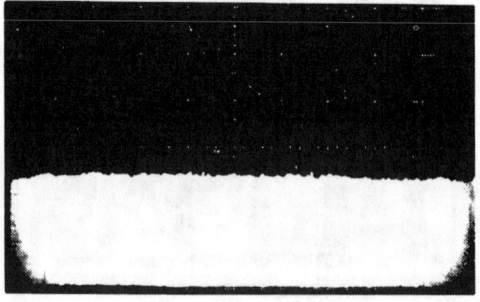

(A) Time Domain

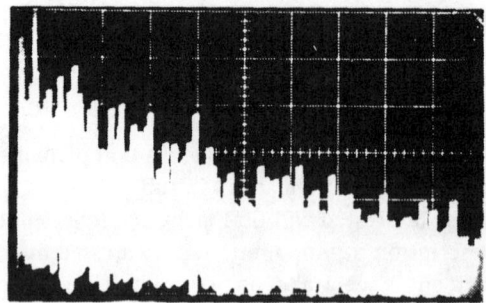

(B) Frequency Domain

Figure 10-7 Zener Noise Source. Bias Setting Gives Unflat Frequency Distribution

---

*Peterson, "Response of Peak Voltmeters to Random Noise," *GR Experimenter,* Vol. 31, No. 7, December, 1956.

Miscellaneous Applications 185

(A) Time Domain

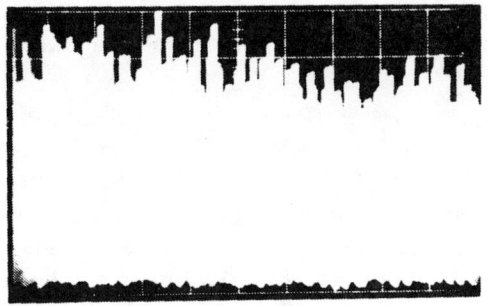

(B) Frequency Domain

Figure 10-8 Zener Noise Source. Bias Setting Gives Flat Frequency Distribution

10.2.2. Absolute Measurements

While absolute noise measurements are difficult, they are not impossible. Here are some of the points that have to be considered.

1. The type of noise, whether random or impulse, has to be identified. Statistically random noise, such as thermal noise, is characterized on a power per unit bandwidth basis. Doubling the spectrum analyzer resolution bandwidth will double the indicated noise power — an increase of 3 dB.

   Impulse noise that is characterized by a succession of narrow pulses, such as ignition noise, is characterized on a voltage per unit bandwidth basis. Doubling the resolution bandwidth will double the indicated noise voltage — an increase of 6 dB. Impulse noise has already been extensively discussed in the sections on pulse analysis though the adjective noise was not used. For example, Equation (5-9) is a mathematical expression of the voltage-bandwidth relationship for such a signal.

   Unless the type of noise being measured is identified, it will be impossible to determine the input power or voltage on a per-unit bandwidth basis.

2. Spectrum analyzer manufacturers usually specify a resolution bandwidth. Depending on the manufacturer, this is either the 3 dB or 6 dB bandwidth. Further, the bandwidth shape factor may be either gaussian or rectangular. These factors can have a great effect on the accuracy of noise measurement. This is because the random noise bandwidth, the impulse bandwidth, the 3 dB bandwidth, and 6 dB bandwidth are seldom equal. Furthermore, they are related to each other differently depending on the bandwidth shape factor.

The effective random noise bandwidth is defined as the normalized width of the power response curve. It is usually obtained by dividing the area (watt-hertz) by the height (watt) on the voltage squared versus frequency response curve. Almost all books on tuned circuits discuss the gaussian amplifier case, where it is shown that the ratio of random noise bandwidth (B_n) to 3 dB bandwidth (B_3) is

$$\frac{B_n}{B_3} = \int_0^\infty \left(\frac{1}{1+y^{2n}}\right)^m dy$$

where m is the number of cascaded stages and n is the number of stagger tuned circuits per stage. For synchronously tuned circuits n = 1 and B_n/B_3 = $\pi/2$ for m = 1, B_n/B_3 = $\pi/4$ (.643) = 1.22 for m = 2. For a perfect rectangular filter, the random noise bandwidth and the 3 dB or 6 dB bandwidth are, of course, equal.

3. Another source of error in noise level measurement relates to spectrum analyzer circuitry; namely, type of detector (linear or square law), detector time constants, and logarithmic amplifier characteristics. A discussion by Sutcliffe* deals with some of the detector problems. He shows that the measurement error depends on the ratio of post-detector to predetector bandwidths. In spectrum analyzer nomenclature, these are the video filter bandwidth (B_V) and resolution bandwidth (B_R). Maximum fractional error is shown to be about $\sqrt{2B_V/B_R}$. Thus, for best accuracy in random noise measurement, the narrowest bandwidth video filter possible should be used. Otherwise, it is difficult to estimate where the peak of the noise really is.

What is usually of interest is to determine the noise power spectral density in watts/Hz, or some equivalent. However, current spectrum analyzers do not use true rms detectors but rather peak detectors. By use of a narrow bandwidth video filter, or other means, the display is that of the noise average rather than rms value. These differ in voltage amplitude by the ratio $(4/\pi)^{1/2} \to 1.05$ dB for a narrowband gaussian process (Rayleigh distribution).

---

*Sutcliffe, "Relative Merits of Quadratic and Linear Detectors in the Direct Measurement of Noise Spectra," *The Radio and Electronic Engineer*, February, 1972.

Miscellaneous Applications

Also, logarithmic display compression reduces the apparent average value. In other words, the logarithm of the mean value is not the same as the mean value of the logarithm. This introduces an additional error of 1.45 dB when operating the spectrum analyzer in the logarithmic vertical mode. The total error is 2.5 dB as shown in the literature.*

Finally it should be noted that the spectrum analyzer produces random noise internally (the sensitivity level). What is displayed is not the incoming noise signal, but rather the sum of input noise and internal noise. Thus, for example, the display will show 3 dB greater than the incoming signal when input noise equals internal noise. This source of error can be ignored for large input noise signal levels (10 dB above internal). Table 10-1 shows appropriate correction factors.

Table 10-1
Combining Two Noise Sources

Measured Noise dB Above Internal Noise	Actual Noise dB Above Internal Noise	Actual Noise dB Below Measured Noise
1.0	-5.87	6.87
1.5	-3.85	5.35
2.0	-2.33	4.33
2.5	-1.09	3.59
3.01	0	3.01
4.0	1.80	2.20
5.0	3.35	1.65
6.0	4.74	1.26
7.0	6.03	0.97
8.0	7.25	0.75
9.0	8.42	0.58
10.0	9.54	0.46

10.2.3. Making the Measurement

First consider whether the total noise power — usually one watt — might be sufficient to damage the spectrum analyzer. Keep in mind that one milliwatt/kHz uniformly distributed over one MHz is one watt; and over one GHz it would be a kilowatt. This is so even though the spectrum analyzer would show one milliwatt when measured with a kHz bandwidth.

The opposite problem is too low a power level. Note that reducing the resolution bandwidth to improve sensitivity will not help for noise signals. The signal (external noise) to noise (internal noise) ratio will not improve as both are equally

---

*Spectrum Analysis — Noise Measurements, Hewlett-Packard App Note 150-4, January 1973.

affected by a change in bandwidth. A low noise high gain preamplifier is the only solution for this problem.

Suppose the noise power level is appropriate for measurement. What then?

1. Choose a convenient resolution bandwidth. Too wide a bandwidth will distort the shape of the noise signal and cause measurement errors. Too narrow a bandwidth increases measurement time. A bandwidth one-third to one-tenth that of the noise signal shape is a good choice.
2. Set the span and reference level to display the signal for convenient viewing.
3. Actuate sufficient smoothing, either video bandwidth reduction and/or digital averaging, to get a smooth display of the average level.
4. Apply the appropriate correction factors. For example:

The noise signal shows as 7.1 μV in the linear display mode using a 1 kHz resolution bandwidth. The actual level is 7.1 x $(4/\pi)^{1/2}$ = 8.0 μV. In the logarithmic mode the signal appears to be -91.4 dBm. Actually it is -91.4 + 2.5 = -88.9 dBm. Both results are the same in a 50 Ω system.

Removing the input signal shows that the self-generated spectrum analyzer noise level is 10 dB (-101.4 dBm) less than the apparent input level. From Table 10-1, note that the input noise is almost 0.5 dB less in amplitude than that observed, or -88.9 - 0.5 = -89.4 dBm.

Manufacturers' literature shows that the noise bandwidth is approximately 0.8 the 6 dB resolution bandwidth. Therefore, 800 Hz is a more accurate number for the bandwidth than 1 kHz. However, resolution bandwidths are specified at ± 20%. An actual measurement of the 6 dB bandwidth takes about five minutes. It shows a 950 Hz bandwidth or a 950 x 0.8 = 760 Hz random noise bandwidth. For highest accuracy, it is best to measure the actual effective random noise bandwidth, which might be off by about 5% from the computed number. Pretend that the random noise bandwidth was measured as 750 Hz. Therefore, the noise level is -89.4 + 10 log 1000/750 $\cong$ -88.2 dBm per kHz.

Another example on using the correction factors:

Wishing to determine the noise properties of a new type of 100 KΩ resistor, connect this resistor across the input of a 7L5/L3 spectrum analyzer, which is set for 1 MΩ input impedance. The result shows -121 dBv (dB with respect to one volt) at 1 kHz bandwidth, as compared to -130 dBv for a shorted input.

The theoretical noise level for a normal resistor is

$$E = \sqrt{4RKTB} = \sqrt{4 \times 10^5 \times 1.37 \times 10^{-23} \times 290 \times 10^3} = 1.26 \, \mu V; -118 \, dBv$$

Miscellaneous Applications

It appears that instead of having a noisier than normal resistor we have one that is 3 dB better than theory predicts. Unfortunately, it is not so. Thus, the noise bandwidth is 750 Hz, the spectrum analyzer will show a noise level 2.5 dB less than actual when in log mode, and the noise will show 0.6 dB greater than the input due to the internal noise that is 9 dB (130 − 121) below the total. The result is −118 − 2.5 + 0.6 − 10 log (1000/750) ≅ −121.1 dBV. The resistor is an ordinary resistor without special noise properties.

10.2.3. EXAMPLES

1. Figures 10-9(A) and 10-9(B) show the noise output from a receiver intermediate frequency (IF) amplifier at 21.4 MHz. From Figure 10-9(A), note that over the 20 MHz full screen frequency span, the noise signal is at least 10 dB greater than the spectrum analyzer internal noise (lower trace). Therefore, internal noise contribution to the total display is negligible. Figure 10-9(B) at 1 dB/DIV shows a 3 dB bandwidth of 3 MHz (6 DIV at 0.5 MHz/DIV). This display was taken at 100 kHz resolution bandwidth, which at one-thirtieth the noise shape width will show the true shape. The widest usable bandwidth would be 1 MHz at a 3:1 ratio, though some shape distortion might occur at that point.

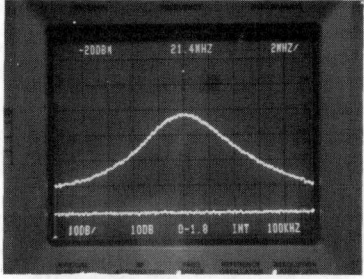

(A) Full Shape Definition

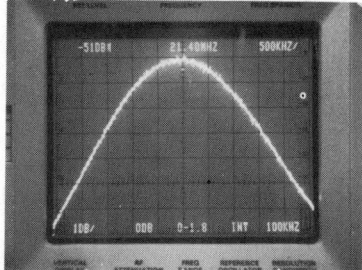

(B) Bandwidth Determination

Figure 10-9 Noise Signal; Shape and Bandwidth Determination

Peak power spectral density shows as −52 dBm at 1 dB below the full screen reference level of −51 dBm. Actual noise level is −52 + 2.5 = 49.5 dBm in an 80 kHz noise bandwidth.

2. Figure 10-10 shows a signal to noise ratio measurement involving phase noise sidebands. The noisy signal to be measured is in the center. The clean signal to the left shows that the noise sideband level generated by the spectrum analyzer is well below that to be measured. At 2 MHz (1 DIV) offset, double sided noise sideband level is −70 dBc (dB below the carrier) in a 100 kHz resolution bandwidth. Actual level is 2.5 dB greater and the noise bandwidth is only 80 kHz. Since $10 \log 80 \times 10^3 = 49$ dB, the noise is −70 + 2.5 − 49 = −116.5 dBC/Hz at 2 MHz offset from the carrier.

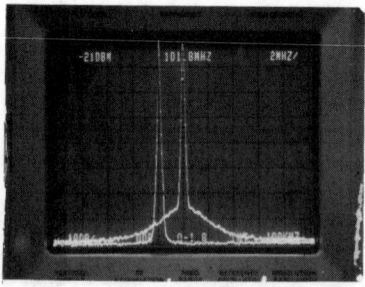

Figure 10-10 Signal to Phase Sideband Noise Determination, Right Hand Signal — Noisy Oscillator; Left Hand Signal — Clean Oscillator

Figure 10-11 shows the close-in noise characteristics of the signal. The clean and noisy signals show identical displays up to about 10 kHz (2 DIV) offset from the carrier. This is the phaselock loop bandwidth of the spectrum analyzer, and the noise seen is added by the spectrum analyzer. Outside the loop bandwidth we see a divergence of the two traces.

Noise closer-in than about 5 kHz cannot be determined as shown in Figure 10-12. Here the noise shows as −73 dBc 100 kHz out in a 10 kHz resolution bandwidth. The usual correction factors should be applied to these numbers.

3. One aspect of signal to noise determination is that the signal and noise will combine to cause erroneous signal level determination when these are of roughly equal amplitude.

We have already discussed the fact that internal noise will combine on a direct power basis with a noise-signal input to show a 3 dB increase when these are equal. A CW signal will not combine with noise on a direct power basis as the spectrum analyzer does not provide a true rms detector. A CW signal when com-

Miscellaneous Applications

bined with noise of equal display level will show a 2.1 dB rather than 3 dB increase. Figure 10-13 and associated example shows the appropriate correction technique when dealing with CW signals combined with noise.

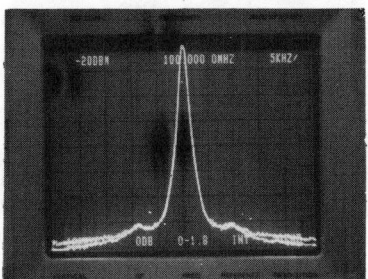

Figure 10-11 Comparing Spectrum Analyzer Generated Phase Noise Sidebands with Noise on the Input Signal Close to the Carrier

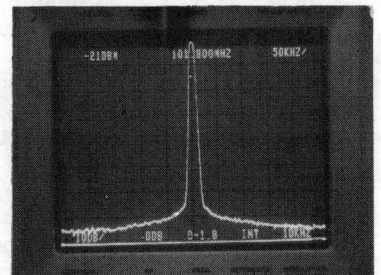

Figure 10-12 Checking Phase Noise at 50 kHz to 250 kHz Frequency Offset

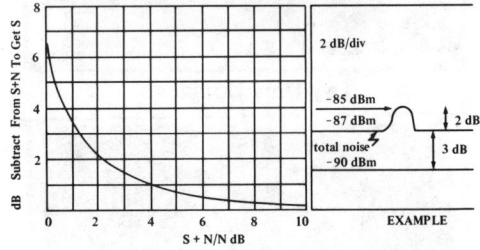

Example: S+N/N ≈ 2 dB hence, S is about 2 dB less than shown by spectrum analyzer. Spectrum analyzer noise is about 3 dB below total noise, hence, input noise is about 3 dB below total noise. In log mode, actual noise is 2.5 dB greater than shown by spectrum analyzer. Therefore:

Actual S/N = 2 − 2 + 3 − 2.5 = +0.5 dB

Actual signal S = −85 − 2 = −87 dBm

Actual noise N = −87 − 3 + 2.5 = −87.5 dBm

S/N = −87 − (−87.5) = +0.5 dBm.

Figure 10-13 CW Signal Plus Noise Correction Factor and Example

10.3. DISTORTION MEASUREMENT

A frequent spectrum analyzer application is to determine the degree of distortion in what is supposed to be a sinusoidal waveform. The method consists of measuring the relative amplitude between the fundamental and the various harmonics and computing the percentage of harmonic content. When there is more than one significant harmonic, the percentages are usually combined by the rms sum method — that is, by taking the square root of the sum of the squares. Figure 10-14 illustrates a typical percentage distortion measurement. Figure 10-14(A) is the oscilloscope presentation of a signal source output; the waveform is obviously not a perfect sinewave. The degree of distortion is, however, very difficult to ascertain from this presentation. Figure 10-14(B) is a spectrum analyzer display for the same waveform. From the spectrum analyzer display, observe that the several harmonics are 25 dB, 33 dB, 38 dB, 50 dB, 42 dB, and 48 dB below the fundamental. Converting from dB to voltage ratios: 1/17.8, 1/44.7, 1/79.4, 1/316, 1/126, and 1/251 are the amplitudes of the harmonics with respect to the fundamental.

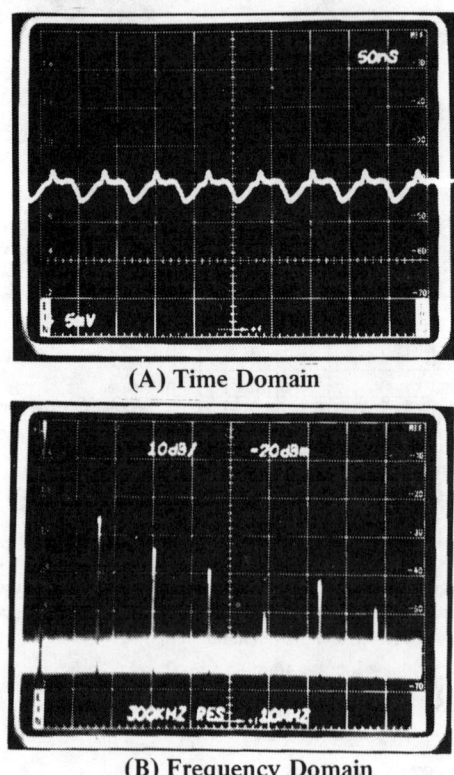

(A) Time Domain

(B) Frequency Domain

Figure 10-14 Distortion Measurement

Miscellaneous Applications

The percentage harmonic content is obtained by multiplying each ratio by 100. Thus, the respective harmonic percentages are: 5.6%, 2.2%, 1.3%, 0.3%, 0.8%, and 0.4%. The total harmonic distortion (HD), computed by the rms sum method, is

$$HD = \sqrt{(5.6)^2 + (2.2)^2 + (1.3)^2 + (0.3)^2 + (0.8)^2 + (0.4)^2} = 6.2\%$$

10.4. COMPONENT TRANSFER – CHARACTERISTIC MEASUREMENT

Basically, a spectrum analyzer is a receiver. As such, it can be used in all applications where a receiver is called for. Thus, the spectrum analyzer can be used as the indicator in VSWR measurements, attenuator-insertion-loss determination, antenna pattern monitoring, filter bandpass and Q determination, etc.

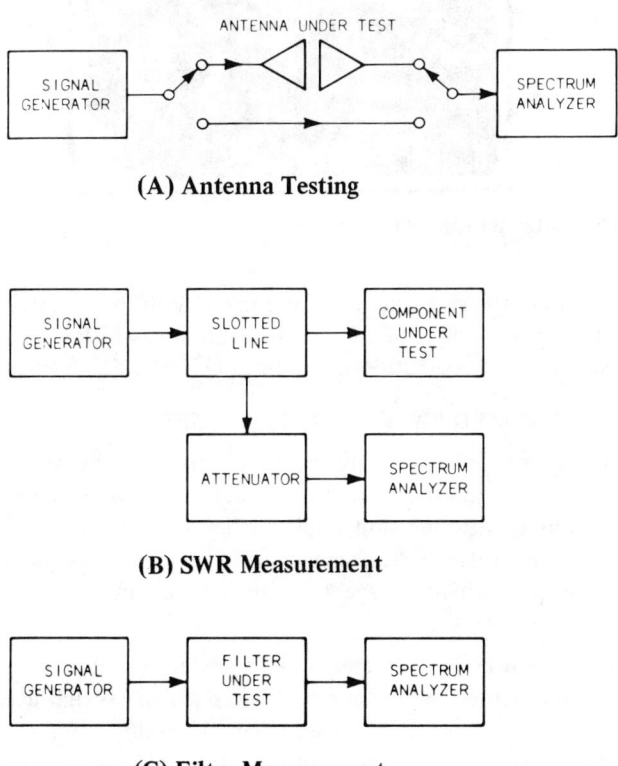

(A) Antenna Testing

(B) SWR Measurement

(C) Filter Measurement

Figure 10-15 Test Setups

The basic test setup consists of a signal generator, a spectrum analyzer, and the component under test. Typical test setups are shown in Figure 10-15. Figure 10-16 illustrates a filter bandpass characteristic and Q measurement. The display was generated by tuning the generator frequency through the spectrum analyzer span. The response will be held in the Max Hold memory for instruments with digital storage. The filter response can be obtained by controlling the shutter speed manually while the signal generator is tuned when digital storage is not available.

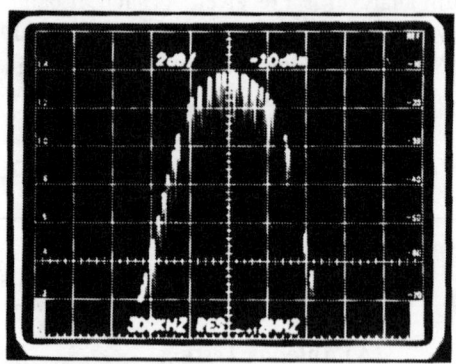

Figure 10-16 Filter Bandpass Characteristic

Figure 10-16 shows the filter characteristic taken with the spectrum analyzer vertical control in the 2 dB/DIV mode. The 3 dB bandwidth is about 5.2 MHz. With a center frequency of 500 MHz the filter Q is $500/5.2 \approx 96$.

10.5. SYNCHRONIZED SWEEPER TECHNIQUES

In a swept front-end spectrum analyzer, the sweeping oscillator frequency setting not only controls the span but also uniquely determines the center frequency. By heterodyning, filtering, and amplitude leveling the sweeping oscillator output, it is possible to generate a signal whose frequency is equal to the instantaneous frequency to which the spectrum analyzer is tuned. Such a device is called a Tracking Generator.

This system can be used to characterize the transfer function characteristics of all types of components. The test methods are the same as that used in the standard sweeper-detector-oscilloscope system. The only difference is that the spectrum analyzer acts as a synchronous detector, thereby providing several advantages over the ordinary peak-detector method. These advantages are:

1. The spectrum analyzer has very high sensitivity, thereby permitting the testing of components that cannot stand the relatively high (over 100 mV) voltages needed in the detector-oscilloscope system.

Miscellaneous Applications   195

2. Synchronous detection filters out the effect of sweeper harmonics, hence, making it easy to accurately characterize components having multiple passbands.

The major application of the tracking generator/spectrum analyzer system is in the testing of filters. Figure 10-17 shows the results of such a measurement.

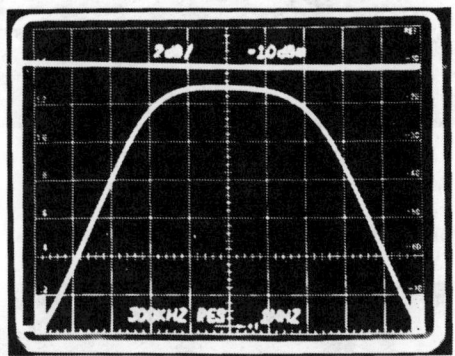

Figure 10-17 Filter Response Displayed on Spectrum Analyzer Using Tracking Generator

From this display, observe the loss in signal transfer when the filter is inserted is about 1.2 dB. The bandwidth at the 3-dB-down points is approximately 5.5 MHz. The loaded Q of the filter can be computed as $Q = f_0/B = 500/5.5 \approx 91$. More detailed skirt selectivity data can be obtained by operating the spectrum analyzer in its logarithmic vertical mode. In short, as with any other sweeper-detector system, this method permits the complete characterization of a filter.

Besides the standard filter measurement applications, the analyzer/tracking generator system can be used in many other areas. One illustration is in the testing of tape recorders. In this application, the output of the TG is connected to the recorder instead of the microphone output, and the playback goes to the spectrum analyzer. This permits the measurement of such things as recorder frequency response and distortion. Figure 10-18 shows the frequency response of a portable tape recorder at the two extremes of the tone adjustment settings. The upper trace shows a 3-dB-down point of about 30 kHz while the lower trace, at a different tone control setting, shows a 3-dB-down point of about 20 kHz. The "birdie" at about 70 kHz is caused by a bias oscillator inside the tape recorder.

Figure 10-19 illustrates the measurement of the distortion characteristics of the tape recorder. Here, a 2 kHz tone was checked for harmonic content straight out of the signal generator (Figure 10-19(A)) and as a playback from the tape recorder (Figure 10-19(B)). It will be observed that going through the recorder has increased the third harmonic by about 10 dB. The tracking frequency was offset, in these measurements, to compensate for the playback time delay.

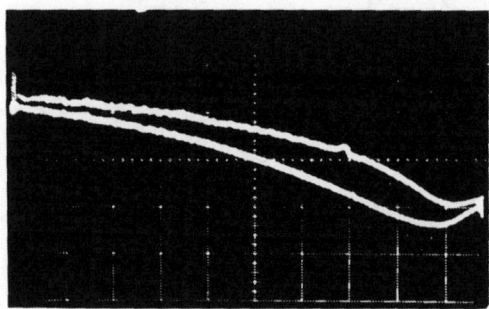

Horizontal: 10 kHz/DIV
Vertical: Log, 10 dB/DIV

Figure 10-18 Portable Tape Recorder Frequency Response

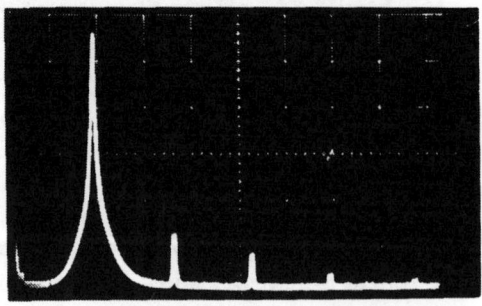

(A) Horizontal: 1 kHz/DIV
Vertical: Log

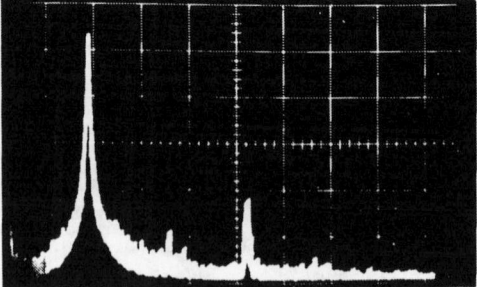

(B) Horizontal: 1 kHz/DIV
Vertical: Log

Figure 10-19 Portable Tape Recorder Distortion Measurement

10.6. EMI MEASUREMENTS

The spectrum analyzer has become an indispensable tool in many areas of the discipline variously known as electromagnetic interference (EMI), radiofrequency interference, or electromagnetic compatibility (EMC). Although the above names and others do not precisely mean the same thing, these are frequently used interchangeably. The subject is a major discipline in its own right. What follows is a very basic introduction to the uses of spectrum analyzers in this area.

All electronic signals, or circuits, have the capacity to interfere with the operation of other circuits or signals or they themselves can be interfered with. Design, installation or maintenance requires, therefore, testing for compatible operation. The residual, internally generated spurious response is a case of an instrument — the spectrum analyzer — interfering with its own proper operation. Such testing requires a receiver such as the spectrum analyzer. Modern spectrum analyzers can do almost everything that specialized EMI receivers can, but not quite. Specialized receivers have multiple detector modes, particular bandwidths or filter shapes to comply with published standards, specialized front-end protection against very wideband signals, specialized readout nomenclature, or special accessories. On the other hand, the sweeping spectrum analyzer makes extremely fast measurements, has many convenience features for design and diagnosis, and is applicable in all but a few specialized cases. Currently the spectrum analyzer and specialized EMI receiver are used side by side, each filling a particular function.

All measurement techniques discussed in this book apply to EMI. The only difference is that the signal is something we actually want to eliminate or minimize. Narrowband EMI signals have the characteristics of other CW signals. Broadband signals are usually impulsive in nature, and the theory and techniques discussed in Chapters 3 and 9 apply. Absolute measurement requires a knowledge of the impulse bandwidth resulting in a voltage spectral density usually in dB above a microvolt per MHz. Radiated measurements are in field strength, i.e., per meter. This requires information on the antenna in use.

Frequently all that is of interest is a relative measurement to determine the degree of change or improvement in a design. Figure 10-20 shows an elementary example involving transformer radiation. The improvement is in the ratio of 20 log 5.5/2 = 8.8 dB. A more advanced measurement would involve determining the absolute rather than relative level of the EMI signal. The next step might involve comparison against a previously established limit, or shielding configuration, or circuit layout, or whatever. Many such elements will be found in the referenced literature.

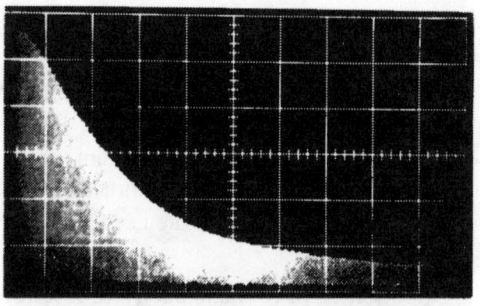

(A) Unshielded Transformer

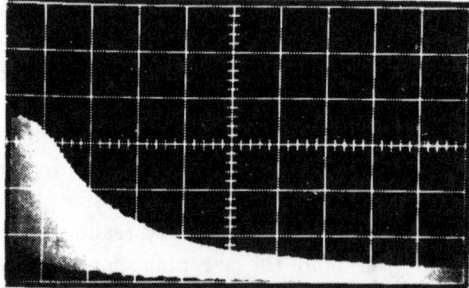

(B) First Shielding Configuration

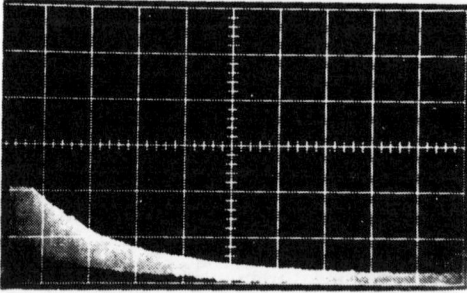

(C) Second Shielding Configuration

Figure 10-20 Shielding Effectiveness Test on Power Transformer. Horizontal 200 Hz/DIV, Center 1 kHz, and Vertical Lin

10.7. TELEMETRY SUBCARRIER TESTS

Many sets of telemetry data are usually transmitted together by modulating a set of subcarriers, which are then used to modulate the main carrier by multiplexing. Subcarrier bands can have either a constant frequency spacing or a proportional frequency spacing. In the proportionally spaced system, the lower frequency bands are spaced closer in frequency than the upper frequency bands. The proportional spacing scheme is illustrated by the table of center frequencies, Table 10-2.

Table 10-2
IRIG Proportional Subcarrier Bands

Band Number	Center Frequency (Hz)
1	400
2	560
3	730
4	960
5	1300
6	1700
7	2300
8	3000
9	3900
10	5400
11	7300
12	10500
13	14500
14	22000
15	30000
16	40000
17	52500
18	70000

Frequently it is desired to check on the frequency spacing, amplitude, and presence or absence of these subcarriers. The spectrum analyzer does an excellent job here, except for the difficulty caused by the crowding of the lower frequency channels. One solution is to use a specialized logarithmically sweeping spectrum analyzer to spread out the CRT spacing of the lower frequency bands. Another solution is to observe the complete set of bands and then expand the region of interest across the full CRT. The latter technique is described in the following.

Figure 10-21(A) shows bands 6, 7, & 8 and 16, 17, & 18. Note that the frequency spacing between the early bands is very close so that a detailed examination is difficult. Figure 10-21(B) shows bands 6, 7, & 8 as these are expanded across the screen.

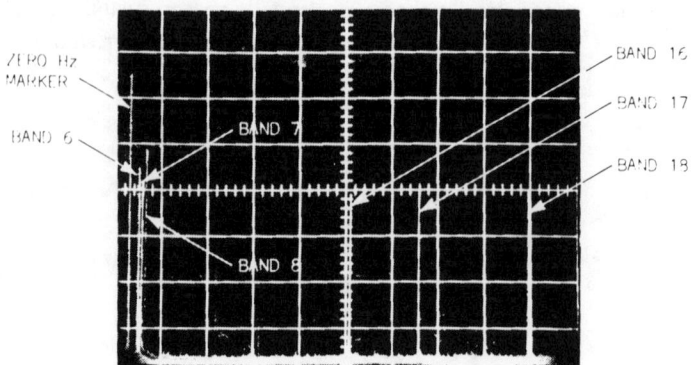

(A) Bands 6, 7, 8, 16, 17, and 18 of Proportional Telemetry Subcarriers

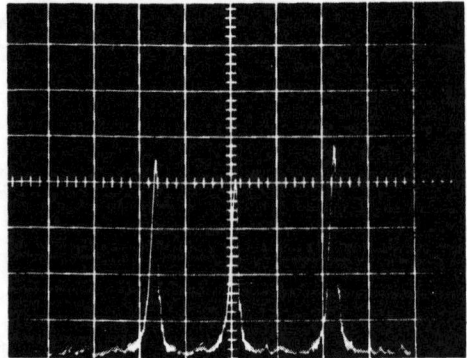

(B) Frequency Around Bands 6, 7, and 8

Figure 10-21 Checking Proportionally Spaced Telemetry Subcarriers by Expansion Method

10.8. DOPPLER VELOCITY MEASUREMENT

The Doppler principle permits the measurement of velocity by bouncing electromagnetic energy off a moving target and measuring the frequency difference between the transmitted and reflected radiation. The best known Doppler velocity measuring system is Doppler radar. For a CW system, the Doppler difference frequency (f_D) is related to the transmitted frequency (f_t), the target velocity (v), and the speed of electronic radiation (c), by $f_D \cong (2v/c)f_t$. For a 100 MHz radar frequency, this represents a Doppler frequency of 29.3 Hz for a 100-mile-per-hour target. These representative numbers indicate that low frequency spectrum analyzers are well suited to this application. Depending on the basic radar frequency, the horizontal scale of the analyzer can be calibrated directly in miles per hour. The target velocity is then read from the horizontal position of the signal on the screen.

Miscellaneous Applications

A more complex Doppler velocity measurement is illustrated by Figure 10-22. Here, the spectrum analyzer is used to measure the Doppler frequency of laser radiation scattered by a moving fluid.

The coherent light of a laser beam is split into two beams of equal path length. The beam directions are so arranged that one beam impinges directly onto a photomultiplier tube while the other beam is pointed away from the photomultiplier, as is shown in Figure 10-22. Some of the light from the main beam, which is pointed away from the photomultiplier, is scattered by the moving fluid and enters the photomultiplier along with the light from the other beam. The photomultiplier acts as a mixer, combining the two light beams and producing an electrical signal at the difference frequency. This difference frequency is determined by the velocity of the fluid, which is, thus, indirectly measurable by the spectrum analyzer.

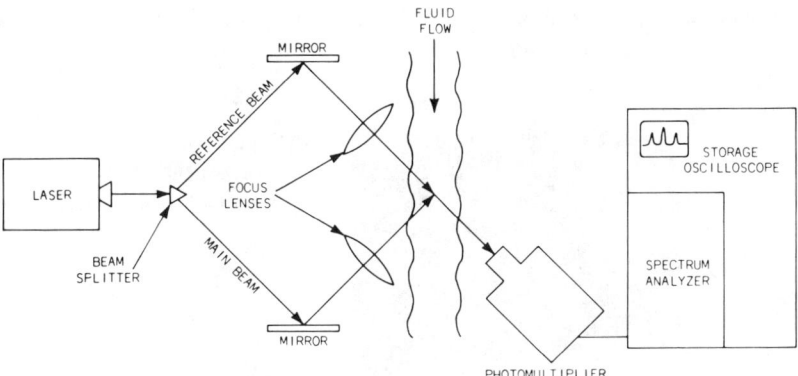

Figure 10-22 Doppler Method of Fluid Velocity Measurement

10.9. USING TRANSDUCERS

Spectrum analyzers can be used as frequency-selective instruments for signals involving transducers. These can involve such diverse applications as cyclical temperature effects, resonant-frequency determination for vibrating bodies, frequency and amount of displacement for various structures, pressure gradients or changes of moving fluids, etc.

Figures 10-23 and 10-24 illustrate the kind of results that one can get. Figure 10-23(A) is a time domain display of the vibration of the floor in a large industrial building under normal usage. The pickup is a velocity transducer having a sensitivity of 600 mV for every inch per second of velocity. From this display, observe that the basic frequency is about $1/10.1 = 10$ Hz and that the peak-to-peak velocity excursion is $200 \cdot 10^{-6}/600 \cdot 10^{-3} = 33$ μin/s. Figure 10-23(B) shows the frequency domain characteristics associated with Figure 10-23(A). The frequency of greatest velocity is clearly closer to 7 Hz than the 10 Hz estimated from the time domain data. The transducer output at this frequency is 400 mV rms. The floor velocity at the 7 Hz resonant frequency is $400/600 \cdot 10^3 = 667$ μin/s rms, or $(667)(2.8) \cong 1870$ μin/s P-P. The spectrum analyzer data shows that, at the resonant frequency, the floor is moving a great deal more than the average time domain data indicates.

Figure 10-24 shows another transducer application. Here, a small mechanical structure is made to vibrate by excitation from a loudspeaker type of exciter, which was in turn driven by a low frequency squarewave. The pickup is a velocity transducer. Clearly, most of the output is at low frequencies. This should be expected, since the driving source (squarewave) has most of its output at the low end of the spectrum. While the low frequency effects are mainly due to the driving waveform, the peaks at 20 kHz, 30 kHz, and 45 kHz are resonance effects in the structure under test.

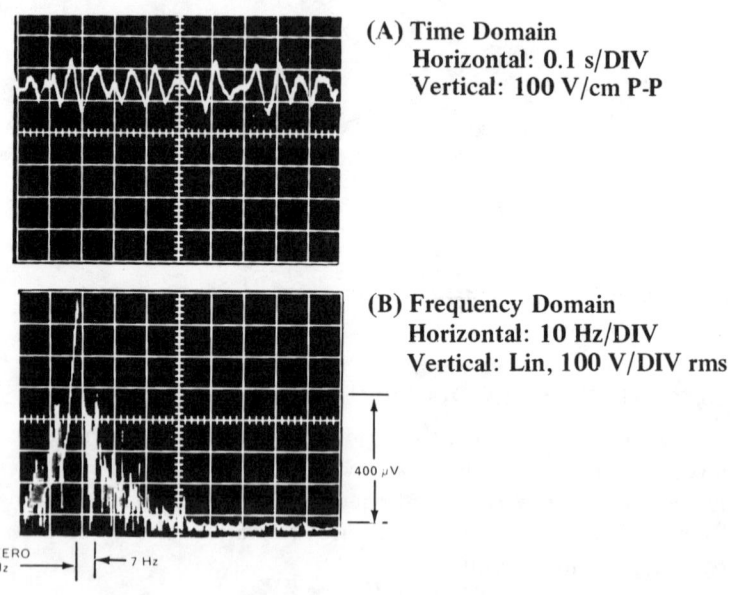

(A) Time Domain
Horizontal: 0.1 s/DIV
Vertical: 100 V/cm P-P

(B) Frequency Domain
Horizontal: 10 Hz/DIV
Vertical: Lin, 100 V/DIV rms

Figure 10-23 Floor Movement — Velocity Transducer, 600 mV for 1-in/s Velocity

Miscellaneous Applications

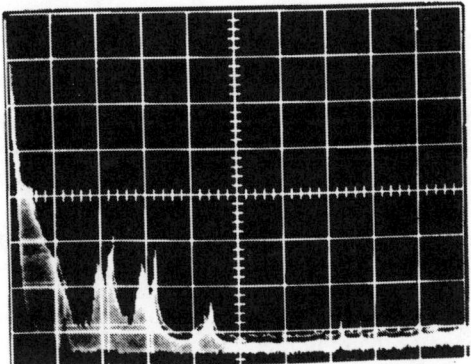

Figure 10-24 Resonance Effects of Small Mechanical Structure as Picked up by Vibration Transducer and Displayed on a Spectrum Analyzer

10.10. SENSITIVITY IMPROVEMENT

Most spectrum analyzers have a very poor noise figure, F. A noise figure of 40 dB or more is not uncommon for microwave spectrum analyzers. A typical front-end might consist of a tracking filter preselector (loss about 6 dB), followed by 5 dB matching circuitry, a very broadband mixer (fundamental conversion loss about 10 dB and noise temperature ratio near unity), a microwave frequency first filter (loss about 4 dB), a narrowband second conversion mixer (loss 4 dB and negligible diode noise), all followed by the first gain at about 6 dB noise figure.

Noise figure and hence sensitivity can be improved by adding low noise figure gain in one or another strategic place. The result can be computed using the basic noise figure relationships summarized below.

Noise figure F is a measure of the degradation in signal to noise ratio (S/N) as the signal passes from the input (i) to the output (o). Thus $F = (S/N)o/(S/N)i$. Noise figure is degraded both by signal loss and an increase in noise above the basic kT_0B from a resistor. For $T = 290°K$, the noise level is -144 dBm at $B = 1$ kHz. For no loss in signal level and no increase in noise, the noise figure is unity, and the sensitivity noise level equals the basic circuit noise, namely -144 dBm at 1 kHz bandwidth. From these basic principles it can be shown that:

$F = L$, the loss of a strictly lossy circuit such as an attenuator.

$F = Lt$, the product of loss and noise temperature ratio for a mixer.

$F = F_1 + (F_2 - 1)/G + (F_3 - 1)/G_1 G_2$ is the relationship between gain and noise figure of circuits in cascade.

$F = L(F_{if} - 1 + t)$ is the noise figure of a mixer followed by an amplifier of noise figure F_{if}.

All of the above are numeric ratios. Noise figure in dB is FdB = 10 log F.

Consider now a spectrum analyzer having a 30 dB (1,000X) noise figure and maximum signal handling level of whatever. Set a zero distortion perfectly flat 20 dB (100X) gain, 10 dB (10X) noise figure amplifier in front. What happens? $F = F_1 + (F_2 - 1)/G = 10 + (999/100) = 19.99 \rightarrow 13$ dB, for a 17 dB improvement. The input signal level must be reduced by 20 dB in order to keep the input to the spectrum analyzer from exceeding the maximum input, whatever that is. Therefore, there is a loss in dynamic range of 20 - 17 = 3 dB.

In real life, the preamplifier will have various limitations so that the loss in dynamic range will be worse than the theoretical example above. Usually a preamplifier will provide a worthwhile improvement in sensitivity when only small signals are involved. When large as well as small signals are involved it may not be worthwhile due to loss of dynamic range.

10.11. DIGITAL RADIO SIGNALS

The last of the Miscellaneous Applications examples deals with phase shift keying (PSK) modulation used in digital radio. This is an excellent example of the great complexity encountered in spectrum analyzer measurements in spite of the fact that this instrument only makes four basic measurements — absolute frequency of one signal component, relative frequency difference between two signal components, absolute amplitude of one signal component, and relative amplitude difference among several signal components.

Usually man-made pulsed digital signals are represented in the frequency domain by an amplitude spectral density. Display amplitude changes in the ratio 20 log bandwidth. Not so for the PSK signal, which behaves as does natural random noise. Display amplitude changes in the ratio 10 log bandwidth. Yet, the basic signal has a clean sin x/x appearance characteristic of the voltage amplitude spectral density due to pulsed RF. Thus, the signal is easily mistaken for a carrier burst involving the use of impulse bandwidth and peak detection (no averaging); while in actuality absolute amplitude measurement calls for signal averaging and the use of random noise correction factors, such as 2.5 dB in the logarithmic display mode.

Many of the basic spectrum analyzer capabilities come into play in characterizing this signal, which is shown in the prefiltered, sin x/x state in Figure 10-25. Checks on signal to noise ratio involve a knowledge of how to process the signal (e.g., peak or average detection) as well as how to correct for measurement bandwidth.

Determination of signal level requires knowledge of how to correct for detection bandwidth and other correction factors based on the type of signal (e.g., 2.5 dB in log mode). Determination of transmission bandwidth, usually 50 dB down

Miscellaneous Applications 205

from mean output level, requires a calculation of the display level drop with respect to the total mean output level. This is followed by a bandwidth measurement the requisite dB down from display peak, as is illustrated in Figure 10-26. The actual bandwidth determination involves various techniques depending on required accuracy and instrument capabilities. For highest accuracy it is usually necessary to use a frequency counter as most spectrum analyzers cannot achieve the better than 1% accuracy required. The latest, computer enhanced spectrum analyzers, such as the Tektronix 494, contain a counter within the instrument so high accuracy is easily attained. Paradoxically these high accuracy instruments also provide the simplest quick estimate capability. Thus, the so-called FCC spectrum shape mask shown in Figure 10-27 can be stored in digital memory, which is provided by these instruments, for later recall and comparison against the signal spectrum, as is illustrated in Figure 10-28.

The above illustrates just some of the complexities encountered in what at first appears a fairly routine signal. Refer to the bibliography for more detailed information.

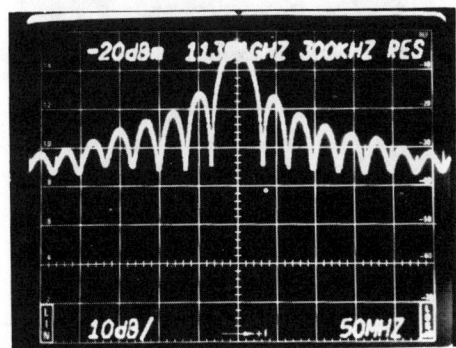

Figure 10-25 Sin x/x Spectrum Generated by 8 PSK Digital Radio Transmission

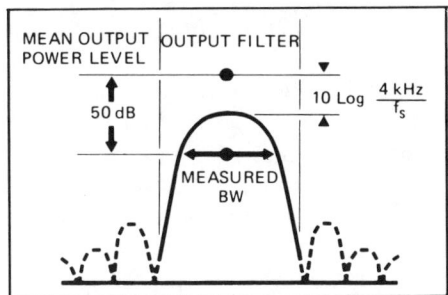

Figure 10-26 Spectrum Output Level Relative to Mean Output Power

206 Modern Spectrum Analyzer Theory and Applications

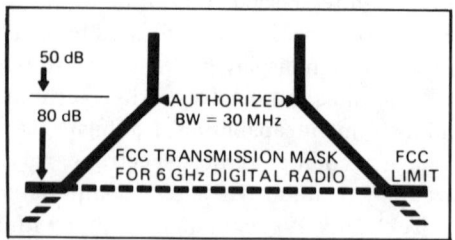

Figure 10-27 Graphical Representation of FCC Transmitted Spectrum Specifications

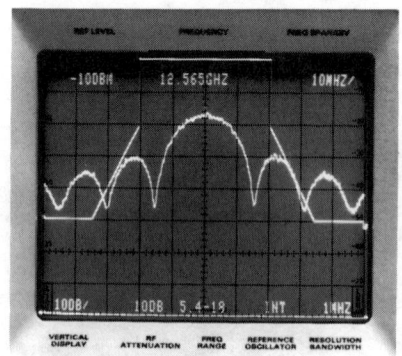

Figure 10-28

10.12. EXERCISES

10-1.
   a. Compute the harmonic distortion for the signal source shown in Figure 10-19(A).
   b. Compute the harmonic distortion for the tape recorder output shown in Figure 10-19(B).
   c. Compare the results of Figures 10-1(A) and 10-1(B).
   d. Assume only the second harmonic is present. How far below the original signal would it have to be for the harmonic distortion to be 0.5%? Give answer in dB.

Note: Assume the vertical display is set at 5 dB/DIV.

Miscellaneous Applications 207

10-2.

    a. Figure 10-25 shows 10 consecutive exposures taken at 1 minute intervals. The nominal frequency is 1 GHz. Compute the warmup drift in PPM/10 minutes.

    b. Could the above described measurement be made if the resolution bandwidth were 3 MHz rather than 30 kHz? Explain.

    c. Reconsider Figure 10-2(B) if the resolution bandwidth were 30 Hz. Explain.

10-3.

    a. You are cited for driving at 60 MPH in a 30 MPH zone by an officer using a police doppler radar. After conversing with the officer, you learn that the radar carrier frequency is 10.5 GHz. What was the recorded doppler frequency as you passed the trap?

    b. Your return to the site with a spectrum analyzer and observe the doppler radar carrier. The carrier shows strong 2 kHz sidebands. Should you contest the citation?

    c. If you observed a 1 kHz beatnote between the doppler radar carrier and the carrier of a nearby microwave relay station, what would be the nominal speed recorded on the doppler radar set?

Chapter 11
Specifications

11.1. INTRODUCTION

In spectrum analyzers, as in other technical areas, there are a great many specialized technical terms. Many people are unfamiliar with the meaning of some of these terms, thus, creating a communications problem. The problem is compounded by the fact that different manufacturers sometimes use different terms to denote the same parameters and, conversely, the same word may have a different meaning. Furthermore, some of the specifications are interrelated so that measurement limitations depend on a combination of several specified parameters.

Most of the above difficulties have been largely eliminated by the publication of various national and international standards. These standards contain both definitions of terms and measurement procedures, as well as other items such as analysis of errors or environmental requirements. "Expression of the Properties of Spectrum Analyzers," IEC publication 714, 1981, provides information on over 50 spectrum analyzer related terms. The U.S. National Standard is "IEEE Standard for Spectrum Analyzers," IEEE Standard 748, 1979. The remainder of this Chapter is devoted to a discussion on some of the more difficult and important terms, especially as it related to measurement techniques and interrelationship between terms.

11.2. SENSITIVITY

Sensitivity is a measure of the smallest level signal that can be detected, observed, or measured. It is an important specification given that an instrument is useless unless signals of interest can be detected. Various elements enter into the specification and determination of sensitivity.

Noise Figure (F). In the absence of special signal processing techniques, the ultimate ability of a receiver, such as a spectrum analyzer, to detect small signals is determined by front-end losses and noise. This is indicated by the noise figure, as is discussed elsewhere in this volume. The equivalent noise level (N), in dBm, is related to noise figure (F), in dB, by $N = -144 + F$ for a 1 kHz noise bandwidth. Noise power is directly proportional to noise bandwidth (B_n) and changes as 10 log Bn.

Sensitivity could be specified by stating the noise figure. However, this is not the usual practice in spectrum analyzers. Among the difficulties of doing it this way is the fact that the specified resolution bandwidth does not equal the random noise bandwidth, and that the spectrum analyzer usually does not indicate the true noise level (see Section 10.2).

S + N = 2N. One of the earliest techniques for specifying sensitivity is the signal plus noise equals twice noise method. With a vertical square law (power) display and rms detector, this is equivalent to a signal power level equal to the noise level. This would be equivalent to the noise figure method if the true noise bandwidth and noise level were known. Unfortunately, this technique has been used under all kinds of conditions; linear (voltage) vertical display, logarithmic vertical display, and differing detector and noise smoothing (averaging) factors. This method is now seldom used.

Equivalent Input Noise. The advent of spectrum analyzers having a calibrated vertical display has led to the almost universal adaptation of the equivalent input noise method. The technique is simply to determine the spectrum analyzer's internally generated noise level referenced to the front-end. No external signal source is necessary, and the result is as accurate as the spectrum analyzer vertical calibration permits. Those looking for improved accuracy should check and/or adjust the vertical reference level of the spectrum analyzer using an accurate signal source or a signal source and power meter.

It should be noted that under ideal conditions all three methods would yield the same result. Thus a 30 dB noise figure and 1 kHz bandwidth yields a −114 dBm sensitivity. A true rms detector would show a doubling of output power level for a −114 dBm signal input, and a calibrated reference level will show −114 dBm of equivalent front-end noise. A real spectrum analyzer of current construction might have an 800 Hz noise bandwidth when the 6 dB resolution bandwidth is 1 kHz; also the displayed noise level will be 2.5 dB below true rms when viewed in the logarithmic vertical mode. Therefore, the true noise level in a 1 kHz noise bandwidth would be 10 log 1/0.8 + 2.5 ≅ 3.5 dB greater than shown, or −114 + 3.5 = −110.5 dBm for a 33.5 dB noise figure. Conversely a 30 dB noise figure would show a −117.5 dBm noise level.

11.3. RESOLUTION

Resolution is for the horizontal (frequency) display what sensitivity is for the vertical (signal level) display. Resolution refers to the ability to distinguish between, to separate, to resolve adjacent in frequency signal components.

Specifications   211

Equal Signals. For equal amplitude signals the measure of resolution is the frequency separation of two display responses, which merge with a 3 dB notch. This definition is illustrated in Figure 11-1(A).

Two equal amplitude signals can be obtained in two ways. One is to use two separate signal generators with their outputs added in a combiner network. Another method is to use a balanced modulator to produce suppressed carrier AM where the two sidebands form the two equal-emplitude signals. Figure 11-1(A) was obtained by use of a balanced modulator. Here the two responses are 1.5 divisions apart. At a span of 200 kHz/DIV, we have 300 kHz.

Usually the use of two signal sources is considered too burdensome a requirement. Hence, the use of a derived specification termed resolution bandwidth. The choice is to specify either the 3 dB or 6 dB bandwidth. The 3 dB bandwidth is the best known of the various choices while the 6 dB bandwidth closely approximates the basic, resolution specification (the actual number for the type of detector used in current instruments is 5.1 dB). Figure 11-1(B) illustrates this definition.

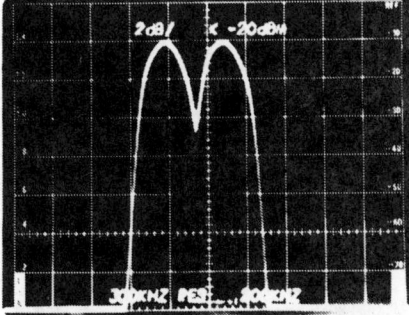

(A) Two Signal Method

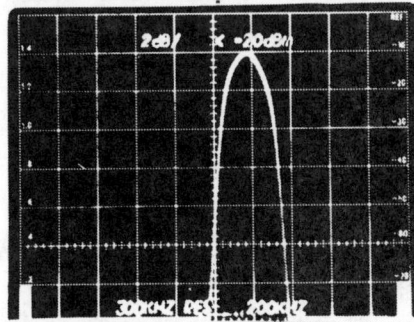

(B) Single Signal Method

Figure 11-1 Measuring Resolution

Unequal Signals. Real life signal components are seldom of equal amplitude. Hence, the need to define skirt resolution, which refers to the signal separation of unequal amplitude signals when the notch formed between display responses is 3 dB down from the smaller.

As with the equal amplitude case, most manufacturers use a derived term known as shape factor. Shape factor is defined as the ratio between the 60 dB bandwidth and the resolution bandwidth. The smaller the shape factor, the closer in frequency can a smaller component be with respect to a large signal component and still be distinguished as separate. See Figure 11-2.

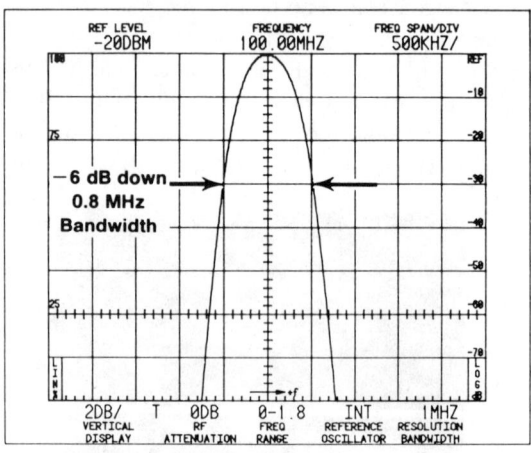

(A) Measuring 6 dB Down Bandwidth

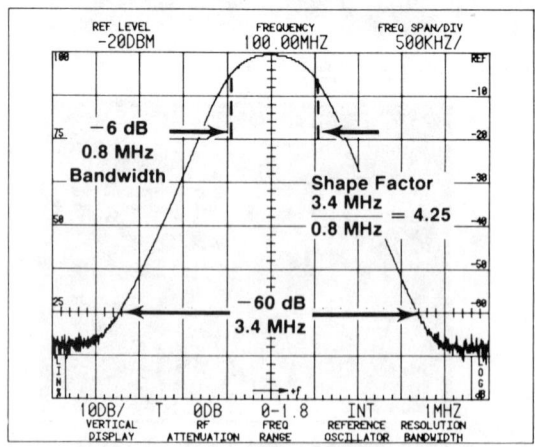

(B) Measuring 60 dB Down Bandwidth and Computing Shape Factor

Figure 11-2 Displays That Illustrate How Bandwidth and Shape Factor are Determined

Specifications 213

Optimum Resolution. For every setting of sweep time and span, there is some particular value of minimum obtainable resolution. Making the resolution-amplifier bandwidth narrower only serves to increase the display width (resolution) observed on the screen. This minimum obtainable resolution is called the optimum resolution (R_o). The fact that an R_o exists is illustrated in Figure 11-3. This is a multiexposure showing the changes in the display as the amplifier resolution bandwidth is changed at constant span and sweep time. As the resolution bandwidth is decreased, the display amplitude decreases and the trace broadens out somewhat. Note that the 3 kHz and 300 kHz bandwidth settings give a wider response width than the 30 kHz setting.

The theoretical optimum resolution can be computed from $R_o \cong \sqrt{S/T}$:

$$R_o = \sqrt{\frac{5 \times 10^4}{10^{-4}}} = \sqrt{5 \times 10^8} = 22 \text{ kHz}$$

This is in reasonably good agreement with the experimental value of 30 kHz.

Optimum resolution should not be confused with the best (narrowest displayed bandwidth) resolution obtainable under normal operating conditions. This is usually limited by the narrowest bandwidth available in the instrument. Optimum resolution is the best that can be done when the span and Time/DIV cannot be changed for some reason.

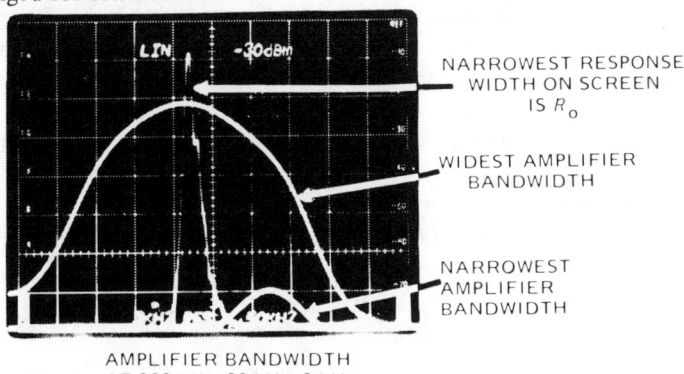

AMPLIFIER BANDWIDTH
AT 300 kHz, 30 kHz, 3 kHz

Figure 11-3 Illustrating Optimum Resolution

Limiting Factors. Normally the limiting factor for equal amplitude signals is the narrowest available bandwidth and the shape factor of that filter. However, instability in local oscillators will frequently prevent the use of the narrowest filters.

Incidental fm should not be greater, peak to peak, than the resolution bandwidth. For a clean-looking display, the ratio should be three to one. For small signals next to large ones, we are also limited by internally generated phase noise sidebands.

Comparing Specifications. It should be clear from the above that comparing specifications between different instruments is not simply a matter of comparing two numbers. Resolution bandwidths cannot be compared directly except where both are defined at the same level (3 dB or 6 dB). The relationship between 3 dB and 6 dB bandwidths is approximately $\sqrt{2}$ for synchronously tuned (Gaussian) filters and a lesser number ($\cong 1.1$) for square type filters. The same caution applies to the comparison of shape factors given that shape factor is the ratio between 60 dB and resolution bandwidths.

When seeking to determine which instrument has the best (narrowest) resolution capability, compare oscillator instability at the frequency of interest as well as available bandwidths. Incidental FM and phase noise sidebands rather than bandwidth are frequently the limiting factors at microwave frequencies.

Other Bandwidths. Usually a precise knowledge of resolution bandwidth is not needed. This is one reason why at ±20%, resolution is among the least accurate specifications.

Power and voltage spectral density measurement, however, does call for accurate bandwidth information. The random noise, or power bandwidth, is needed when determining noise level in watts/Hz, and the impulse bandwidth is a necessary factor in establishing spectral density in volts/Hz for various pulse type signals including electromagnetic interference due to impulsive signals. Neither of these bandwidths equals the resolution bandwidth, either 3 dB or 6 dB. Approximate correction factors are usually published by manufacturers. For best accuracy, it is necessary to measure each filter separately. Measurement techniques are discussed in the sections on random noise and pulse measurement.

11.4. STABILITY

Stability is of interest in two ways. In one respect, the spectrum analyzer is a tool used to measure the frequency stability of incoming signals. A second aspect is to specify and determine the stability, or more accurately instability, of the spectrum analyzer.

Three parameters are usually of interest: drift, incidental fm, and phase noise sidebands. Line related coherent sidebands is frequently specified separately but is actually a subset of incidental fm.

Drift is basically a slow, long-term phenomenon measured in seconds, minutes, or hours. The industry appears to have settled on 10 minutes as the basic time interval, although Hz/min and Hz/hr is also used. Drift is especially affected by warm up time. Thus, a drift specification of one kHz/10 min two hours after turn-on does not necessarily imply better drift performance than 10 kHz/min measured 30 minutes after warm up. Most manufacturers will provide a single

Specifications

guaranteed number, usually one hour after warm up, and supplemental typical performance as a function of time since instrument turn on. Another important factor is the outside, ambient environment. Drift measurement is simply a matter of checking the display frequency change associated with a known to be stable signal. The above discussion is not applicable to the newest sweep and lock spectrum analyzers where long-term drift is for all intents and purposes zero.

Incidental fm refers to fast relatively small frequency excursions as opposed to the slower drift. Figure 11-4 illustrates the difference between drift and incidental fm. The time duration for measuring incidental fm is on the order of seconds or milliseconds. Figure 11-5 illustrates the effect of measurement time. The fm is caused by 60 Hz and 120 Hz power supply ripple. At 20 mS/DIV (A), it is not possible to observe any trace broadening due to incidental fm. At 500 μs/DIV (B), the fm is clearly observed. At 10 S/DIV (C), the fm deviation appears to have increased. Actually we are observing a combination of incidental fm and drift.

The manufacturer will usually specify the appropriate measurement time. Various means for measuring fm deviation, whether incidental or deliberate, is discussed elsewhere in this volume. One of the simplest and most accurate techniques is to use the resolution filter slope as an fm discriminator as illustrated in Figure 11-6. One point to remember: incidental fm is specified on a peak-to-peak basis whereas deviation of an fm signal is usually specified on a zero-to-peak basis.

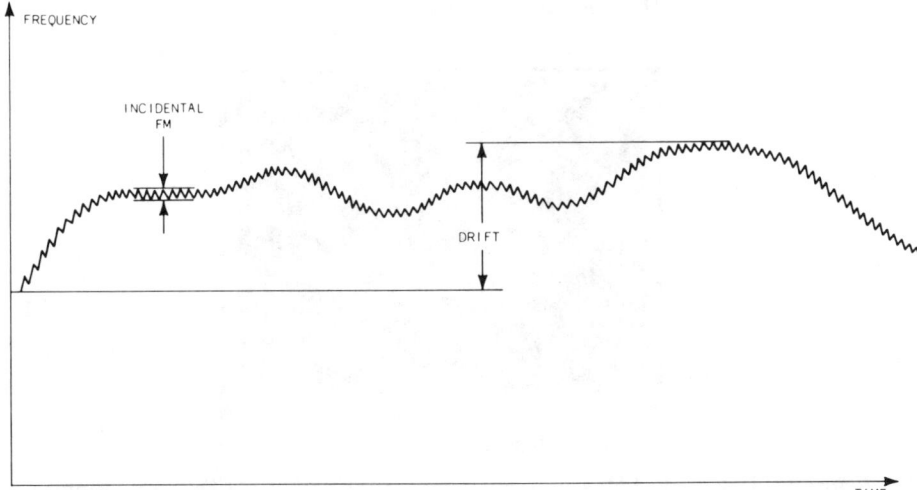

Figure 11-4 Difference in Excursion Period Between Drift and Incidental FM

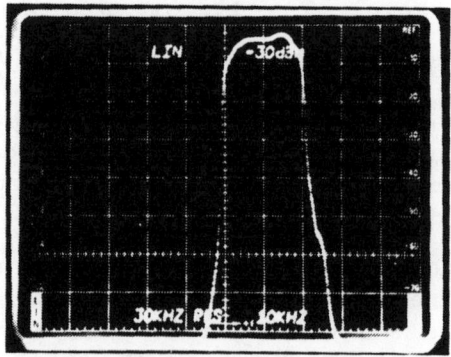

(A) Fast Measurement

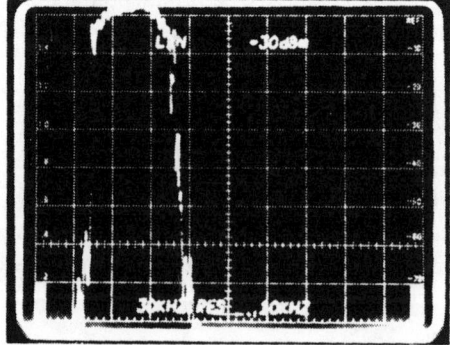

(B) Correct Measurement

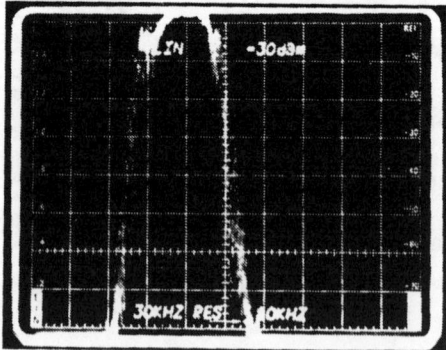

(C) Slow Measurement

Figure 11-5 Effect of Measurement Time on Incidental FM Determination

Specifications

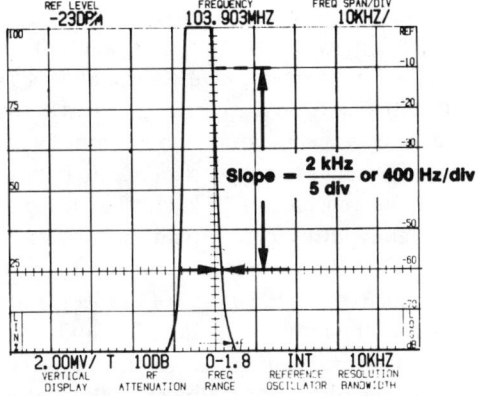

(A) Calculating Slope of Response

(B) Measuring FM as the Deviation/Division of the Response

Figure 11-6 Typical Display Showing How to Determine Residual FM

Phase noise sidebands refer to the low level random noise, on both sides of an oscillator output signal, due to perturbations caused by noise in the oscillator circuit. The noise can be both am and fm and, since the spectrum analyzer will respond to both, noise sidebands is a more accurate term. However, since the phase component tends to dominate, and many measurement techniques eliminate the am, it is usual to use the term phase noise.

Noise sidebands extend over a significant frequency range, diminishing with frequency offset from the carrier signal. Eventually the sideband noise level will drop below the spectrum analyzer sensitivity. A noise sideband specification requires three elements: the frequency separation from the carrier (frequency offset), the measurement bandwidth (resolution setting) and the level below carrier amplitude (dBC). A simple specification provides one number such as: 70 dBC minimum for 50 kHz offset at 1 kHz BW.

More detailed specifications provide information for all bandwidths and varying offset. Such a specification gives a dBC number for an offset equal to a certain number of bandwidths. For example, 75 dBC Min at 30X bandwidth offset. In other words, 75 dBC 30 kHz from the carrier in a 1 kHz BW or 900 Hz offset for a 30 Hz BW, etc. Such a specification, when properly interpreted, can be used to determine performance for any offset at any bandwidth. For example, what performance can be expected at 15 kHz offset using a 1 kHz BW? We know that a 500 Hz bandwidth should provide at least 75 dBC at 15 kHz offset since 30 x 0.5 = 15. We also know that random noise goes 10 log BW ratio. Therefore, for a 1 kHz BW we have 10 log 1000/500 = 3 dB more noise than for a 500 Hz bandwidth and the result is 72 dBC. The above violates basic oscillator noise theory which calls for sideband noise to change at 9 dB/Octave close to the oscillator and 6 dB/Octave further out. However, the method is in line with the logic of the specification. Basically the original specification has considerable safety margin at some offsets. It is not necessarily so for the derived numbers.

A factor that users need to be aware of is that sideband noise is frequently specified only for fundamental conversion. Microwave spectrum analyzers use multiples of the first local oscillator to cover the frequency range. Sideband noise will degrade at the rate 20 log harmonic number (N). Thus, if the specification is 80 dBC for fundamental conversion, it is 74 dBC for second harmonic conversion (20 log 2 = 6 dB).

Finally, it should be noted that the spectrum analyzer shows double sideband noise (as opposed to single sideband), both phase and amplitude. Furthermore, the specification is based on displayed noise level not the true noise level, which

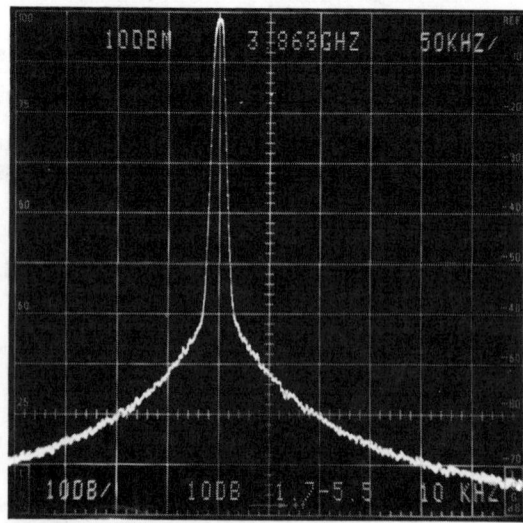

Figure 11-7 Noise Sidebands Illustration. The Photo Shows 70 dBc at 50 kHz Offset, 80 dBc at 100 kHz Offset and 90 dBc at 200 kHz Offset; All at 10 kHz Resolution Bandwidth

Specifications   219

has to be corrected for detector and display factors (2.5 dB in log mode) and noise to resolution bandwidth ratio. Figure 11-7 illustrates spectrum analyzer sideband noise behavior.

Power line frequency related sidebands is a special form of incidental fm. Sideband amplitude is easily related to incidental fm peak to peak deviation through the narrowband fm relationship (Figure 4-25). Thus, 60 Hz sidebands 40 dBC represent a 100:1 ratio → t/2 or t = 1/50 = $\Delta f/fm$ → Δf = 60/50 = 1.2 Hz, and the peak to peak incidental fm is 2.4 Hz.

11.5. DISPLAY FLATNESS/FREQUENCY RESPONSE

Many factors contribute to amplitude measurement error. A significant contribution comes from display unflatness or frequency response. These terms are sometimes used interchangeably, but they are not the same.

Display flatness — more accurately unflatness — is a measure of the uniformity of amplitude response over a specified frequency span. Spectrum analyzer controls are not changed once the initial setting is made. Only the input signal frequency is changed. Frequency response, on the other hand, refers to the uniformity of amplitude response for a signal at center frequency while the input frequency setting of the spectrum analyzer is changed over a specified range. Thus, spectrum analyzer control setting is changed during frequency response measurement. While it is not directly stated in the definition, it is usual practice to optimize other controls, such as preselector peaking, along with center frequency change. At frequencies below a gigahertz or two, there is virtually no difference between the two specifications. At higher frequencies, however, there is a significant impact due to amplitude peaking adjustments. It is not unusual to have a ±1 dB of difference between the two techniques. For example: Display flatness 5 GHz to 12 GHz of 6 dB peak to peak (or ±3 dB) and a frequency response of 4 dB peak to peak (±2 dB).

Another factor of confusion is between peak to peak and plus-minus designation and where the frequency reference point might be designated. A peak to peak designation provides accurate information for relative amplitude accuracy. Thus, two signals of identical amplitude might appear to differ by 4 dB when measured with an instrument specified at 4 dB peak to peak. However, we have no way of knowing whether the smaller appearing signal is shown 4 dB less than its true amplitude, the larger 4 dB greater than its true amplitude, or a combination of both. Therefore, there is a 4 dB relative amplitude error and a 4 dB absolute amplitude error for each of the signals individually. This need not be so. A ±2 dB designation still means a 4 dB relative error. However, the absolute error for each signal is reduced to 2 dB provided the reference point for the plus and minus designation is set at the true (calibrated) reference level. Such a designation requires a trail back to the calibrator frequency since the calibrator is the amplitude setting standard for the whole instrument.

11.6. DYNAMIC RANGE

There are two categories of dynamic range — display and measurement. The display dynamic range depends primarily on the type of logarithmic amplified or other signal compression and display circuits employed. A typical number is 80 dB based on 8 vertical divisions at 10 dB/DIV.

Measurement of dynamic range depends on the kind of signal being measured. Hence, each spectrum analyzer has several dynamic ranges. In each case, the measurement dynamic range is taken as the ratio, or dB difference, between the largest and smallest signal levels that can be intercepted and measured together. The smallest signal level is usually equated to the sensitivity, which leaves us to define the largest level. While several answers are possible, it is generally taken that the largest permissible signal component is one that can be measured and which does not prevent the smallest signal from being measured. Here is how this definition applies to intermodulation distortion.

Intermodulation Distortion (IM): IM occurs when new signal components not originally present in the input signal are generated at the sum and difference frequencies of input signal components and their harmonics. Thus, if the input signal consists of two frequency components, f_1 and f_2, we have IM at $f_1 + f_2$, $f_2 - f_1$, $2f_2 - f_1$, $2f_1 - f_2$, etc. The most disturbing IM components are those that occur near the input signals, namely $2f_2 - f_1$ and $2f_1 - f_2$. This is because these close-in components are difficult to filter out. Intermodulation is identified by an order number equal to the harmonics creating the new signal. Given the above discussion, it should be clear why most interest centers on the third order components.

The usual IM specification will state how far down the undesired components will be when two CW signals of specified amplitude are applied to the spectrum analyzer. For example: third order IM is at least 75 dBC from two -30 dBm signals at any spacing. The matter of spacing is important because IM gets worse as signal level increases. The closer the spacing the greater the possibility that multiple components will simultaneously be present down the IF channels after passing through gain and filtering. Therefore, the distortion tends to increase as signal spacing gets closer. Some specifications will indicate a signal spacing below which the specified numbers no longer apply.

Going back to the example, the specification indicates that there is a 75 dB measurement range for -30 dBm input signals. Suppose we had another instrument specified at 85 dBC for -40 dBm signals. Which is the better dynamic range? Also, what if we wanted to use -35 dBm signals? The easiest way to handle such comparisons is via the intercept point (I).

The intercept point is that signal level where the distortion component amplitude would equal that of the direct signal. The intercept point is usually not reached

Specifications

as the circuit saturates before that point. Figure 11-8 illustrates this concept. The input signal has a one to one input-output correspondence in dB. Of course, the absolute level in dBm is offset by the gain or attenuation of the circuit. The intermodulation component has a slope, on the dB scale, equal to the order number. Thus, the output level for third order components increases at three times the dB rate as the input and the difference changes at three minus one, or two times the input dB rate. The relationship illustrated by Figure 11-8 can be expressed mathematically as follows: $I = dBC/(n-1) + S$ where I is the intercept point in dBm, dBC is the dynamic range dB difference between distortion and normal signal components, n is the order number and S is the signal level in dBm.

We can now answer the questions raised previously. For $dBC = 75$ dB when $S = -30$ dBm, $I = [75/(3-1)] - 30 = +7.5$ dBm. For $dBC = 85$ dB at $S = -40$ dBm, we get $I = +2.5$ dBm. The former yields a better dynamic range than the latter. At -30 dBm, the dynamic range of the second system is only 65 dB → $[2.5 = (dBC/2) - 30]$.

Perhaps one of the more useful concepts is that of optimum dynamic range. It is the best (largest) dynamic range that can be achieved and is obtained at a signal level which causes distortion components equal to the sensitivity noise level. With S_0 equal the optimum signal level that yields dBC_0, the optimum (best) dynamic range we have: $S_0 = [(n-1)I + n]/n$ and $dBC_0 = [(n-1)(I-N)]/n$ where N is the dBm sensitivity noise level.

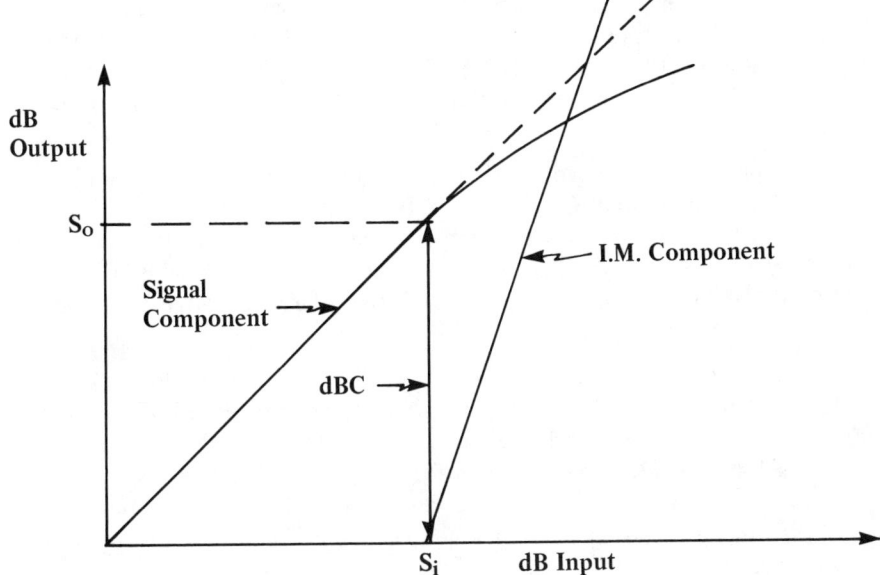

Figure 11-8 Definition of Intercept Point

11.7. COMBINING SPECIFICATIONS

Most specifications are interrelated in fairly complicated ways. Consider the following illustration.

Phase Noise Sidebands: 70 dBC at 20 x BW, fundamental conversion.

Sensitivity at 1 kHz is −120 dBm for fundamental conversion up to 4 GHz and −115 dBm for 2nd harmonic conversion between 4 GHz and 7 GHz.

Third order IM is 80 dBC for −30 dBm signals.

What is the best dynamic range for measuring small signals within 40 kHz of large signals at 6 GHz?

The third order intercept point is $I = (80/2) - 30 = +10$ dBm.

Sensitivity at 6 GHz for a 1 kHz bandwidth is −115 dBm.

Noise sidebands degrade at 20 log n or 6 dB for second harmonic conversion, yielding $70 - 6 = 64$ dBC at 20 x BW.

20 x BW requires a 2 kHz bandwidth for 40 kHz offset. Hence, a 1 kHz bandwidth will show $10 \log 2 = 3$ dB less noise, or $64 + 3 = 67$ dBC at 40 kHz offset and 1 kHz bandwidth.

Phase noise dominates over noise figure, sensitivity, noise for input signal levels above −48 dBm ($48 + 67 = 115$). The dynamic range is limited to 67 dB by the phase noise of the local oscillators.

The best (optimum) dynamic range permitted by the sensitivity is $dBC_0 = [(n-1)(I-N)]/n = \{(3-1)[10-(-115)]\}/3 = 83.3$ dB. This is obtained at an input signal level of $S_0 = [(n-1)I+N]/n = [(3-1)(10)-115]/3 = -31.6$ dBm. The originally specified condition of 80 dBC for −30 dBm signals is near optimum.

However, phase noise sidebands limit the dynamic range to 67 dB at 40 kHz offset. The specified 80 dBC cannot be obtained this close to the carrier. The input signal level need not be kept near −30 dBm. From $I = dBC/(n-1) + S$ we have $10 = (67/2) + S \rightarrow S = -23.5$ dBm. As long as the input signal is less than −23.5 dBm and more than $-115 + 67 = -48$ dBm, the 67 dBC dynamic range will be available.

Additional complications are possible. An unbelievably poor resolution bandwidth shape factor of 100:1 would prevent observation of signals 67 dBC within 40 kHz of the carrier. We might be interested in determining dynamic range for other resolution bandwidth settings, or for other signals (e.g., harmonics rather than intermodulation), or the specification may be written in a different way (e.g., the gain compression point may be specified instead of IM intercept point), etc.

Specifications

The important point to remember is that the various specifications are interrelated. Some work may be necessary in order to get at factors relevant to the measurement of interest.

Appendix

dB, dBm

The logarithmic expression of ratios is quite common in spectrum analyzer usage. The following is a brief review of the exponential function and, its inverse, the logarithmic function, notation in bels and decibels and finally a set of tables of decibel relationships.

An exponential is a relationship of the form $a^m = P$. Some of the rules for the manipulation of exponentials are:

$$(a^m)(a^n) = a^{(m+n)} \tag{A-1}$$

$$\frac{a^m}{a^n} = a^{(m-n)} \tag{A-2}$$

$$a^{-m} = \frac{1}{a^m} \tag{A-3}$$

$$a^{m/n} = \sqrt[n]{a^m} \tag{A-4}$$

$$a^0 = 1 \tag{A-5}$$

The inverse of an exponential function is a logarithmic function. Thus, the relationship $a^m = P$ can also be written as

$$\log_a P = m \tag{A-6}$$

where a is the base and m is the logarithm of P to the base a. Our interest is in a base of a = 10. This is called the System of Common Logarithms. Common logarithms are usually abbreviated as log — this abbreviation is supposed to indicate that the base is 10. Thus, log 100 = 2, since $10^2 = 100$.

The advantage of handling ratios in terms of logarithms is that multiplication and division are eliminated and are replaced by addition and subtraction. This follows from the relationship between exponents of exponential functions as given in Equations (A-1) through (A-5). Our interest is in ratios of power. In logarithmic notation, a power ratio is given as $\log(P_2/P_1) = N$ bels. The bel is a dimensionless unit, simply indicating a power ratio expressed in logarithms to the base 10. The bel is a fairly large ratio, a more convenient unit is one tenth of a bel or a decibel (dB). There are 10 dB in every bel. A power ratio expressed in dB's is

$$10 \log\left(\frac{P_2}{P_1}\right) = N \text{ dB} \tag{A-7}$$

That is, the basic power ratio in bels is multiplied by ten to get the number of decibels. In Equation (A-7), P_1 is the reference power to which P_2 is compared. The result is in the form of a ratio. Thus, for $P_2/P_1 = 100$, $N_{bels} = 2$, and $N_{dB} = 20$. Whether $P_1 = 1$ W with $P_2 = 100$ W, or $P_1 = 10$ W with $P_2 = 1$ kW makes no difference; N is 20 dB if $P_2/P_1 = 100$.

The relationship in Equation (A-7) is the basic definition for decibels, all other relationships are derived from this. Sometimes a derivation is based on assumptions or approximations that will not always hold. When in doubt, the user should go back to Equation (A-7). An example is when dealing with voltage ratios across equal resistances. Thus, $P_1 = V_1^2/R$, $P_2 = V_2^2/R$, and

$$10 \log\left(\frac{P_2}{P_1}\right) = 10 \log \left(\frac{V_2}{V_1}\right)^2 = N \text{ dB}$$

since the resistances cancel. Following the rules for manipulating exponents and logarithms, we get

$$10 \log \left(\frac{V_2}{V_1}\right)^2 = 20 \log \left(\frac{V_2}{V_1}\right) = N \text{ dB} \tag{A-8}$$

Equation (A-8) is only valid when dealing with a constant impedance, otherwise it is incorrect.

Sometimes expressing a specific amount of power using the dB notation is desired. This requires that the reference level P_1 be fixed. When the units are given as dBm, it means that $P_1 = 1$ milliwatt. Thus, 20 dBm means 100 mW. Frequently, a power level less than 1 mW needs to be expressed logarithmically. Here, the ratios are inverted and a minus sign added before the dBm, in accordance with the rules for exponentials, Equation (A-3). Thus,

$$10 \log \left(\frac{.001}{1 \text{ mW}}\right) = -10 \log\left(\frac{1000}{1 \text{ mW}}\right) = -30 \text{ dBm}$$

Table A-1 is a table of voltage and power ratios versus the equivalent number of dB's. The user should keep in mind that the voltage- or current-ratio method only holds true across equal impedances.

Appendix

Table A-1
Decibels

dB	Voltage or Current Ratio	Power Ratio	dB	Voltage or Current Ratio	Power Ratio
0.0	1.000	1.000	26.	19.95	398.1
0.1	1.012	1.023	27.	22.39	501.2
0.2	1.023	1.047	28.	25.12	631.0
0.3	1.035	1.072	29.	28.18	794.3
0.4	1.047	1.096	30.	31.62	1000.
0.5	1.059	1.122	31.	35.48	1259.
0.6	1.072	1.148	32.	39.81	1585.
0.8	1.096	1.202	33.	44.67	1995.
1.0	1.122	1.259	34.	50.12	2512.
1.5	1.189	1.413	35.	56.23	3162.
2.0	1.259	1.585	36.	63.10	3981.
2.5	1.334	1.778	37.	70.79	5012.
3.0	1.413	1.995	38.	79.43	6310.
4.	1.585	2.512	39.	89.13	7943.
5.	1.778	3.162	40.	100.0	10000.
6.	1.995	3.981	41.	112.2	12590.
7.	2.239	5.012	42.	125.9	15850.
8.	2.512	6.310	43.	141.3	19950.
9.	2.818	7.943	44.	158.5	25120.
10.	3.162	10.000	45.	177.8	31620.
11.	3.548	12.59	46.	199.5	39810.
12.	3.981	15.85	47.	223.9	50120.
13.	4.467	19.95	48.	251.2	63100.
14.	5.012	25.12	49.	281.8	79430.
15.	5.623	31.62	50.	316.2	100000.
16.	6.310	39.81	51.	354.8	125900.
17.	7.079	50.12	52.	398.1	158500.
18.	7.943	63.10	53.	446.7	199500.
19.	8.913	79.43	54.	501.2	251200.
20.	10.000	100.00	55.	562.3	316200.
21.	11.22	125.9	56.	631.0	398100.
22.	12.59	158.5	57.	707.9	501200.
23.	14.13	199.5	58.	794.3	631000.
24.	15.85	251.2	59.	891.3	794300.
25.	17.78	316.2	60.	1000.0	1000000.

Table A-2
Carrier Nulls,
$J_0(t) = 0$

Null Number	$t = \Delta F/f$
1st	2.4048
2nd	5.5201
3rd	8.637
4th	11.7915
5th	14.9309
6th	18.0711
7th	21.2116
8th	24.3525
9th	27.4935
10th	30.6346

Table A-3
First-Sideband Nulls,
$J_1(t) = 0$

Null Number	$t = \Delta F/f$
1st	3.83
2nd	7.02
3rd	10.17
4th	13.32
5th	16.47
6th	19.62
7th	22.76
8th	25.90
9th	29.05

BESSEL FUNCTIONS

Bessel functions are used extensively in frequency modulation, as discussed in Chapter 4. Our interest is restricted to Bessel functions of the first kind, integer order, and positive argument. The notation is $J_p(t)$, where: J means Bessel function of the first kind, p is the order, and t is the argument. In frequency modulation theory, the order indicates the sideband number and the argument is the modulation index ($\Delta F/f$). For more detailed tables and graphs of Bessel functions, see the references. For example, *British Association for the Advancement of Science Mathematical Tables,* University Press, Cambridge, vols. VI & X.

Appendix

Table A-4
Bessel Function Values

Argument t = ΔF/f	Carrier $J_0(t)$	First Sideband $J_1(t)$	Second Sideband $J_2(t)$	Third Sideband $J_3(t)$
0.0	1.000	0.000	0.000	0.000
0.2	0.990	0.100	0.005	0.0002
0.4	0.960	0.196	0.020	0.001
0.6	0.912	0.287	0.044	0.004
0.8	0.846	0.369	0.076	0.010
1.0	0.765	0.440	0.115	0.020
1.2	0.671	0.498	0.159	0.033
1.4	0.567	0.542	0.207	0.051
1.6	0.455	0.570	0.257	0.073
1.8	0.340	0.582	0.306	0.099
2.0	0.224	0.577	0.353	0.129
2.2	0.110	0.556	0.395	0.162
2.4	0.003	0.520	0.431	0.198
2.6	−0.097	0.471	0.459	0.235
2.8	−0.185	0.410	0.478	0.273
3.0	−0.260	0.339	0.486	0.309
3.2	−0.320	0.261	0.484	0.343
3.4	−0.364	0.179	0.470	0.373
3.6	−0.392	0.096	0.445	0.399
3.8	−0.403	0.013	0.409	0.418
4.0	−0.397	−0.066	0.364	0.430

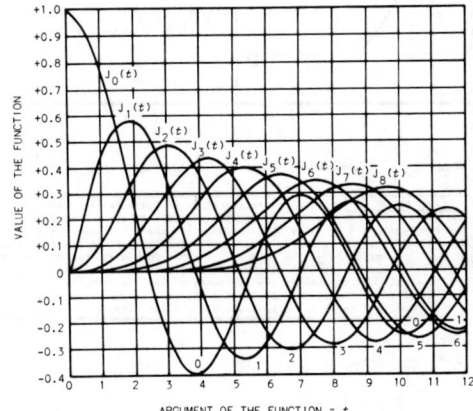

Figure A-1 Bessel Functions for the First Eight Orders

Table A-5
Fourier Transform

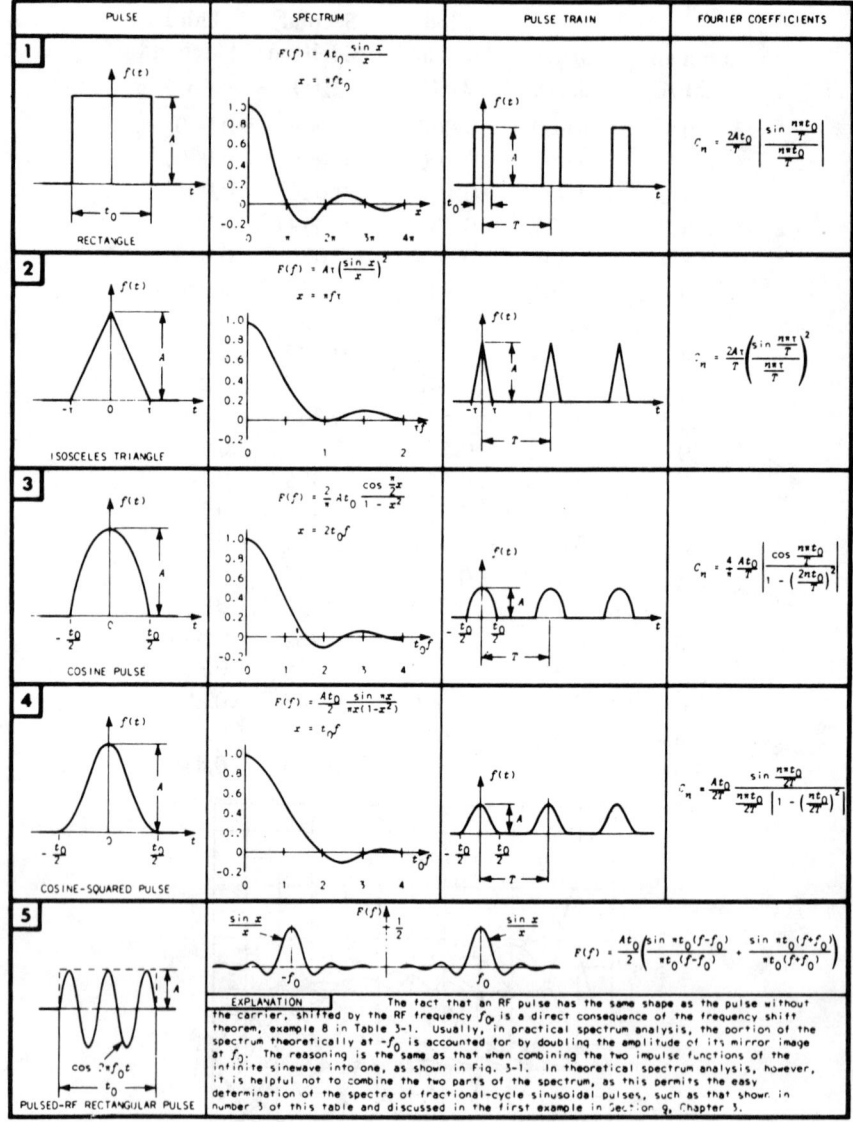

Appendix

Table A-6

	DC Pulse	RF Pulse
Discrete line spectrum (volts rms)	$C_n = \dfrac{2}{\sqrt{2}} \dfrac{At_0}{T} \dfrac{\sin\left(\dfrac{n\pi t_0}{T}\right)}{\left(\dfrac{n\pi t_0}{T}\right)}$	$C_n = \dfrac{At_0}{\sqrt{2}\,T} \dfrac{\sin\left(\dfrac{n\pi t_0}{T}\right)}{\left(\dfrac{n\pi t_0}{T}\right)}$
Dense Continuous Spectrum (volts rms/ Hz of impulse bandwidth, B_i	$S(\omega) = \dfrac{2At_0 B_i}{\sqrt{2}} \dfrac{\sin(\pi f t_0)}{\pi f t_0}$	$S(\omega) = \dfrac{At_0 B_i}{\sqrt{2}} \left[\dfrac{\sin \pi t_0 (f - f_0)}{\pi t_0 (f - f_0)}\right]$

FOURIER ANALYSIS

Table A-5 gives both graphical and mathematical relationships for time domain to frequency domain conversion.

CW SENSITIVITY

The noise power generated by a resistance is $N = kTB$, where N is noise power, k is Boltzman's constant, T is absolute temperature in degrees Kelvin, and B is the noise bandwidth. At an absolute temperature of $290°K$, this is equivalent to a noise power of -114 dBm for a 1 MHz bandwidth. The actual sensitivity of an amplifier is always worse than this because of higher noise due to the active elements and because of signal losses in matching attenuators, filters or other front-end devices. The amount by which the actual sensitivity is degraded compared to the ideal sensitivity is called the noise figure. Thus, the CW sensitivity, as measured by the signal-equals-noise method, is a function of both noise figure and noise bandwidth — the temperature is usually assumed to be fixed at $290°K = 17°C$.

Figure A-2 is a graphical representation of the sensitivity, bandwidth, noise figure relationship.

As indicated in Chapter 5, the CW sensitivity degrades as sweep time and/or resolution bandwidth is decreased. The equation connecting these parameters for a Gaussian resolution shape is

$$\alpha = \left[1 + 0.195 \left(\dfrac{S}{TB^2}\right)^2\right]^{-1/4} \tag{A-9}$$

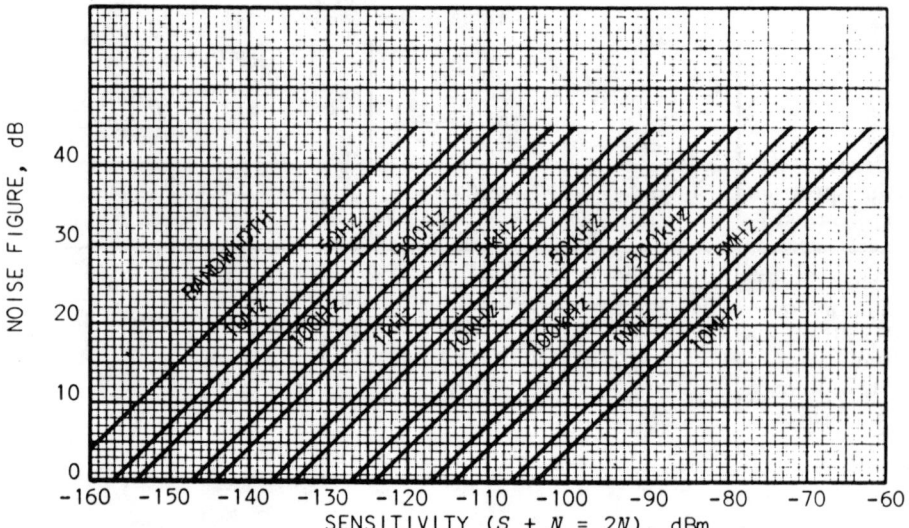

Figure A-2 Sensitivity as a Function of Bandwidth and Noise Figure Noise Level Sensitivity dBm

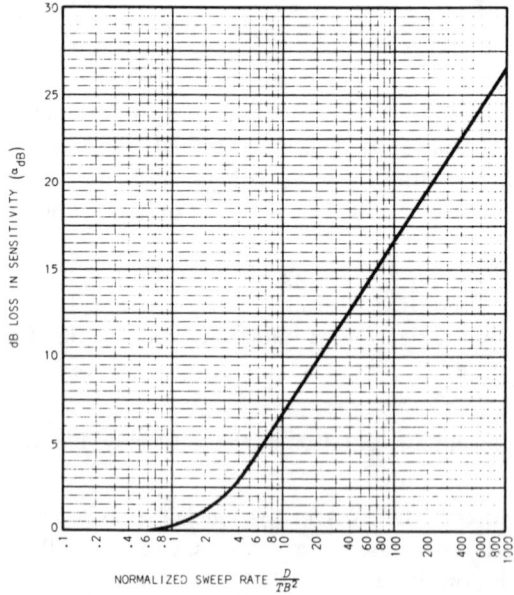

Figure A-3 Loss in CW Sensitivity as a Function of Normalized Sweep Rate. S = span (Hz); B = 3 dB Bandwidth (Hz); T = Time (S); Based on Assumption of Gaussian Amplitude Response dB Loss in Sensitivity (α_{dB})

Appendix

Alpha is the sensitivity ratio, a number less than unity. A convenient way of indicating this loss in sensitivity is in dB, which is computed

$$\alpha_{dB} = 20 \log \alpha \qquad (A\text{-}10)$$

Since α is less than unity, α_{dB} is a negative number. However, when plotting on a graph, the negative sign is usually replaced by the word "loss," which is what the negative sign meant in the first place. Figure A-3 is a graph of sensitivity loss as a function of full-screen span, S, resolution bandwidth, B, and full-screen sweep time T.

RESOLUTION BANDWIDTH

In Chapter 5, it was indicated that there are two terms associated with the resolution capability of the spectrum analyzer. One term is resolution bandwidth (B); this is actual bandwidth of the narrowest bandwidth amplifier. A second term is resolution (R), which refers to the display on the CRT screen. At long sweep times, the display on the CRT is a tracing of the response characteristic of the spectrum analyzer passband and the two terms become synonymous. At short sweep times, the display on the screen indicates a wider bandwidth than the amplifier actually has. The ratio of the screen display (R) and the true bandwidth can be computed from the relationship:

$$\frac{R}{B} = \left[1 + 0.195 \left(\frac{S}{TB^2}\right)^2\right]^{1/2} \qquad (A\text{-}11)$$

This equation is plotted in Figure A-4.

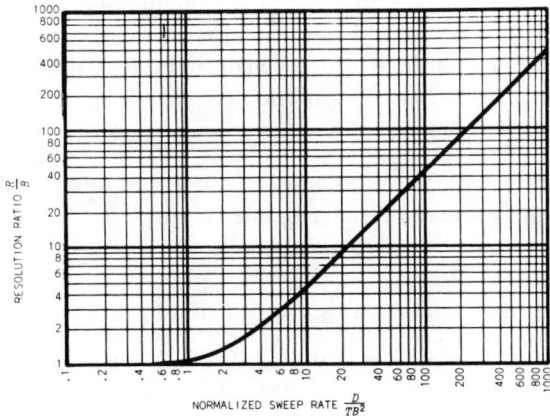

Figure A-4 Loss in Resolution as a Function of Normalized Sweep Rate; S = span (HzO; R = Resolution (Hz); B = Bandwidth (Hz); T = Time (s); Based on a Gaussian Amplitude Response

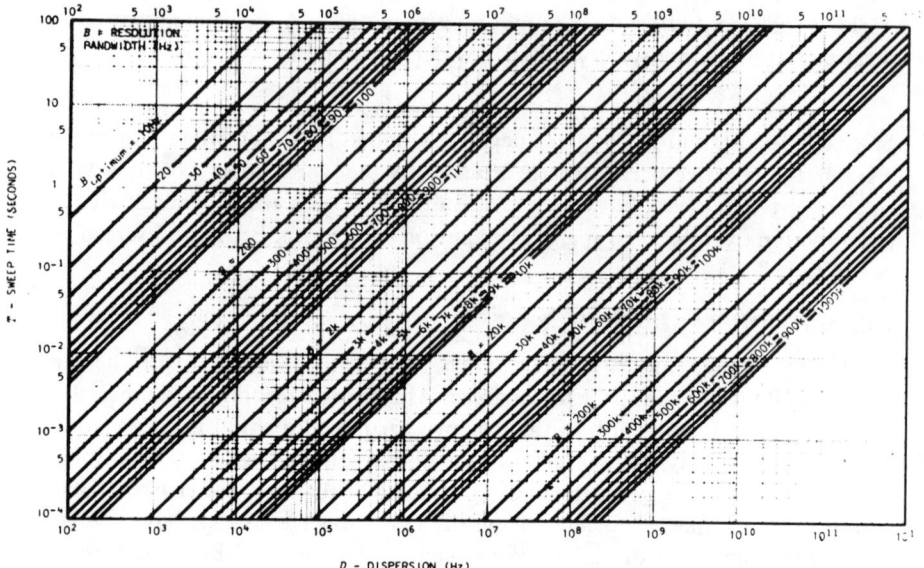

Figure A-5 Optimum Resolution Setting for Spectrum Analyzers. Read Values of $B_{optimum}$ for a Given Dispersion and Sweep Time

The concepts of optimum resolution and optimum resolution bandwidth are discussed in Chapter 5. Basically, at fixed span and sweep time there is one resolution-bandwidth setting which yields the narrowest resolution shape on the CRT display. These are called optimum resolution bandwidth and resolution respectively. Figure A-5 is a graph showing the optimum resolution bandwidth (B_0) as a function of sweep time (T) and span (S). The optimum resolution (R_0), which is the closest spacing of two signals that can be separated on the CRT screen, is related to optimum resolution bandwidth by

$$R_0 = \sqrt{2} \; B_0 \tag{A-12}$$

It should be noted that Figure A-5 is based on the assumption of a Gaussian amplitude response for the resolution amplifier.

PULSED RF

The details on pulsed RF measurements are discussed in Chapter 9.

In order to display the fine details of the spectrum, it is necessary that

$$t_0 B \leqslant 0.1 \tag{A-13}$$

where t_0 is pulse width and B is resolution bandwidth. This relationship is plotted in Figure A-6. When the pulsewidth-bandwidth product is greater than one tenth, spectrum shape details may be lost. As the pulsewidth-bandwidth product gets smaller, there is a progressive loss in sensitivity compared to a CW signal. Thus, $t_0 B = 0.1$ is the ideal setting for pulsed RF.

Appendix

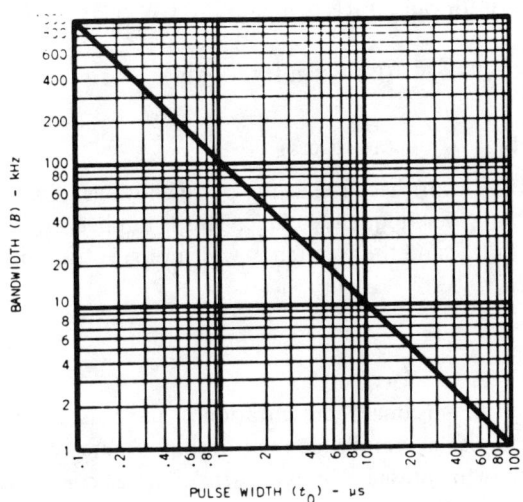

Figure A-6 Resolution Bandwidth Setting for Pulsed RF Computed From $Bt_0 = 0.1$

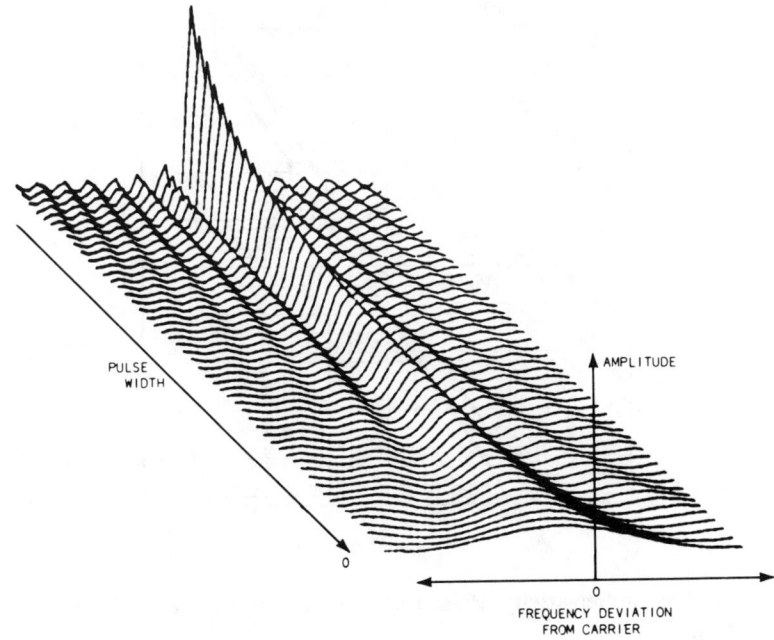

Figure A-7 Pulse Width Sidelobe Width Pulse Sensitivity Relationship

The loss in sensitivity for pulsed RF compared to CW can be computed from:

$$\alpha_{dB} = 20 \log \frac{3}{2} t_0 B_3 \quad \text{(Gaussian filter only)} \tag{A-14}$$

$$\alpha_{dB} = 20 \log t_0 B_i \quad \text{(all filters)}$$

The loss in sensitivity occurs because, as the pulse width gets narrower, the energy spreads out over a wider frequency range. This is implied in the relationship between pulse width (t_0) and the frequency width of spectrum nulls (ΔF), namely,

$$\Delta F = \frac{1}{t_0} \tag{A-15}$$

Figure A-7 is a three-dimensional representation of the pulse width, sidelobe-frequency width and display-amplitude relationship. Figure A-8 is a graph of Equation (A-14), showing pulsed RF loss in sensitivity as compared to a CW signal.

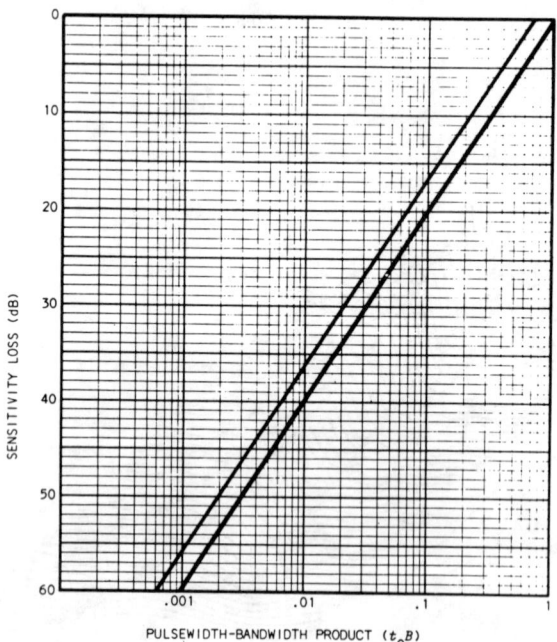

Figure A-8 Loss in Sensitivity, Pulsed RF Versus CW. Upper $\alpha = 1.5\ t_0 B_3$; Lower = $t_0 B_i$

Appendix

RANDOM NOISE

Table A-7
Combining Two Noise Sources

Measured Noise dB Above Internal Noise	Actual Noise dB Above Internal Noise	Actual Noise dB Below Measured Noise
1.0	−5.87	6.87
1.5	−3.85	5.35
2.0	−2.33	4.33
2.5	−1.09	3.59
3.01	0	3.01
4.0	1.80	2.20
5.0	3.35	1.65
6.0	4.74	1.26
7.0	6.03	0.97
8.0	7.25	0.75
9.0	8.42	0.58
10.0	9.54	0.46

Table A-8
Absolute Level Measurement

Linear vertical mode: Multiply by $\sqrt{4/\pi}$

Logarithmic vertical mode: Add 2.5 dB

The above are with respect to the averaged display level using peak detection.

Absolute level changes with bandwidth at 10 log Bandwidth ratio.

Power spectral density is in units of watts/Hz of random noise bandwidth.

For true shape definition use a resolution bandwidth no greater than one-third that of the noise signal.

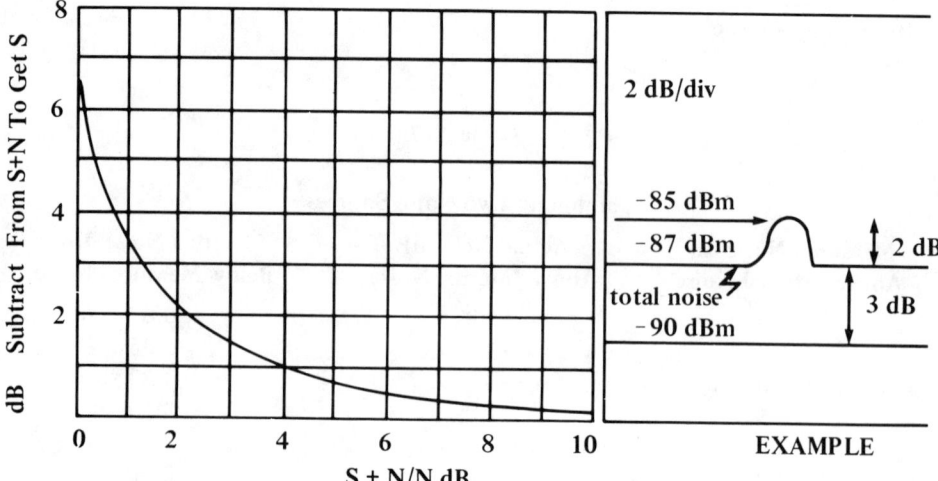

Example: S+N/N ≈ 2 dB hence, S is about 2 dB less than shown by spectrum analyzer. Spectrum analyzer noise is about 3 dB below total noise, hence, input noise is about 3 dB below total noise. In log mode, actual noise is 2.5 dB greater than shown by spectrum analyzer. Therefore:

Actual S/N = 2 − 2 + 3 − 2.5 = +0.5 dB

Actual signal S = −85 − 2 = −87 dBm

Actual noise N = −87 − 3 + 2.5 = −87.5 dBm

S/N = −87 − (−87.5) = +0.5 dBm.

Figure A-9

INTERMODULATION DYNAMIC RANGE

$$I = \frac{dBC}{n-1} + S \qquad (A\text{-}16)$$

$$S_0 = \frac{(n-1)I + N}{n} \qquad (A\text{-}17)$$

$$dBC_0 = \frac{(n-1)(I-N)}{n} \qquad (A\text{-}18)$$

Appendix

SYMBOLS

Symbol	Description	first appears in chapter
a	instantaneous amplitude	4
$a_0, a_1, \ldots a_{m,n}$	Fourier series amplitude constants	2
A	amplitude of waveform	2
A_c	carrier amplitude	7
A_s	sideband amplitude	7
$b_1, \ldots b_{m,n}$	Fourier series amplitude constants	2
B	amplitude of waveform,	2
	filter bandwidth (Hz)	1
B_0	optimum resolution bandwidth (Hz)	5
B_3	3 dB bandwidth (Hz)	10
B_i	impulse bandwidth (Hz)	9
B_n	random noise bandwidth (Hz)	10
B_v	video filter bandwidth (Hz)	10
c	speed of electromagnetic radiation	10
C	capacitance (F)	2
C_o	carrier amplitude in Fourier Series (v)	9
C_n	combined amplitude of nth harmonic in Fourier series	3
d	symbol for differentiation	2
d_n	amplitude of nth harmonic in complex notation of Fourier series	3
D	dispersion (Hz),	5
f	frequency (Hz)	1
	modulation frequency (Hz)	4
f_d	dial frequency (Hz)	6
f_D	Doppler difference frequency	10
f_i	image input frequency (Hz)	6
f_{IF}	IF amplifier center frequency (Hz)	1
f_{LO}	local-oscillator frequency (Hz)	1
f_{mo}	mixer output frequency (Hz)	6
f_p	pulse repetition frequency (Hz), the same as PRF or PRR	9

f_{RF}	signal input frequency (Hz)	1
f_t	transmitted frequency (Hz)	10
f_0	center frequency (Hz)	1
f( )	function of ( )	2
ΔF	frequency offset from carrier (Hz), the same as $f - f_0$	9
F	carrier frequency, noise figure ratio	1, 10
$F(\omega)$	Fourier transform	3
F_{if}	Noise figure of IF amplifier	10
I	intercept point (dBm)	11
j	$\sqrt{-1}$	2
$J_p(t)$	Bessel function of the first kind of order p and argument t	4
k	Boltzman's constant	5
K	amplitude ratio in AM	4
L	inductance (H)	2
m	degree of modulation,	4
	harmonic number	2
n	distortion order number,	11
	harmonic number	2
	intermodulation order number	11
	number of filters	1
N	dBm	11
	noise power (W),	5
Q	charge (C)	2
R	resistance (Ω),	2
	resolution observed on CRT (Hz)	5
R_0	optimum resolution observed on CRT (Hz)	5
$R(\omega)$	real part of complex spectrum	9
S	span or frequency span (Hz)	5
S_0	signal level (dBm) for optimum (best) intermodulation dynamic range dBC	11
$S_0(\omega)$	mainlobe peak spectral density (v/Hz)	9
$S(\omega)$	spectral density shown by analyzer	5
Si(x)	sine integral of x	3

Appendix

t	noise temperature ratio,	10
	time duration (s)	2
t_0	pulse width (s)	3
T	absolute temperature (°K),	5
	period of waveform (s),	2
	sweep time per full CRT width (s)	1
v	velocity	10
x	unknown variable,	2
	variable angle (rad)	3
$X(\omega)$	imaginary part of complex spectrum	9
y	unknown variable	2
Z	complex number	2
Z*	conjugate of Z	2
α	constant phase angle (rad),	2
	loss factor	5
β	modulation index	8
$\delta(t)$	impulse function	3
∂	symbol for partial derivative	2
ΔF	bandwidth,	3
	frequency deviation	4
ϵ	base of natural logarithms, 2.718 ...	2
θ	variable phase angle (rad)	2
π	3.141 ...	2
ρ	amount of error	2
$\sum_{n}^{m}$	summation between the limits n to m	2
τ	system response pulse-width for CW input (s),	1
	time shift interval (s)	3
ϕ_n	phase of nth harmonic in Fourier series (rad)	3
ω	radian or angular frequency, angular velocity	2
=	equals	1
$\neq$	is not equal to	2
$\approx$	is approximately equal to	4
>	is greater than	3
$\geq$	is greater than or equal to	3

$<$	is less than	2
$\leq$	is less than or equal to	5
$\rightarrow$	approaches,	3
	leads to, results in	9
$\Leftrightarrow$	is equivalent by transformation	5
$\uparrow$	increase	6
$\downarrow$	decrease	6
$\sqrt{}$	square root	2
∞	infinity	3
$\|\ \|$	absolute value of quantity within the bars	2
$\lim_{x \to 0}$	indicating the limit of a term as x approaches 0	3
$\int_a^b$	definite integral between limits a and b	2

BIBLIOGRAPHY

General References

Abramowitz and Stegun, *Handbook of Mathematical Functions.* New York: Dover Publications, 1964.

Jahnke and Emde, *Tables of Functions.* New York: Dover Publications, 1945.

Mumford and Scheibe, *Noise Performance Factors in Communication Systems.* Dedham, MA: Horizon House, 1968.

British Association for the Advancement of Science, *Mathematical Tables,* University Press, Vols. VI and X.

"Expression of the Properties of Spectrum Analyzers," *IEC* Publication no. 714, 1981.

IEEE, "Standard for Spectrum Analyzers," *IEEE Standard* Publication no. 748, 1979.
Reference Data for Radio Engineers, ITT.

Spectrum Analysis Theory

Bracewell, *The Fourier Transform and Its Applications.* New York: McGraw-Hill, 1965.

Cushman, "Spectrum Analyzer Myths," Report no. NADC-EL-6452, February 1965.

Goldman, *Frequency Analysis Modulation and Noise.* New York: McGraw-Hill, 1948.

Guillemin, "The Fourier Integral — A Basic Introduction," IRE, *Trans. CT,* September 1955.

Guillemin, *The Mathematics of Circuit Analysis.* London/New York: Wiley and Sons, 1949.

Hartley, "A More Symmetrical Fourier Analysis Applied to Transmission Problems," *Proc. IRE,* March 1942.

Lathi, *An Introduction to Random Signals and Communication Theory,* Scranton International, 1968.

Marcus, "The Significance of Negative Frequencies in Spectrum Analysis," IEEE, *Trans. EMC,* December 1967.

Papoulis, *The Fourier Integral and Its Applications.* New York: McGraw-Hill, 1962.

Stuart, *An Introduction to Fourier Analysis,* Methuen and Company, 1961.

Weber, *Linear Transient Analysis,* Vols. I and II. London/New York: Wiley and Sons, 1956.

Whittaker and Watson, *Modern Analysis.* London: Cambridge University Press, 1958.

The Spectrum Analyzer

Alm, "Microprocessor Simplifies Spectrum Analysis," Tektronix No. A-3722.

Barnard, "New Spectrum Analyzers Bring High Performance to On-Site Measurements," *Defense Systems Review,* February 1984.

Barnard, "Spectrum Analyzer is Portable Lab Assistant," *Microwaves,* January 1984.

Batten, *et al.,* "The Response of a Panoramic Receiver to CW and Pulsed Signals," *Proc. IRE,* June 1954.

Benedict, "Fundamentals of Spectrum Analysis," Tektronix no. 26W-5360.

Bryant and Smith, "Obtaining Accurate Center Frequency in a Portable Spectrum Analyzer," *Microwave Journal,* February 1984.

Chang, "On the Filter Problem of the Power Spectrum Analyzer," *Proc. IRE,* August 1954.

David and Bryant, "Internal Processing Simplifies Automated Spectrum Analysis," *Microwave Journal,* May 1980.

Engelson, "Interpreting Spectrum Analyzer Displays," *Microwaves,* January 1966.

Engelson, "Make the Analyzer Work for You," *Microwaves,* May 1971.

Engelson and Long, "Optimizing Spectrum Analyzer Resolution," *Microwaves,* December 1965.

Engelson, "Reference Level Control in Spectrum Analyzers: Simplification Through Internal Intelligence," *R.F. Design,* March/April 1984.

Engelson, "Spectrum Analyzer Circuits," *Tektronix Circuit Concept Series* no. 062-1055-00, 1969.

Feigenbaum, "Introduction to Spectrum Analyzers," *Microwaves,* April 1963.

Frisch and Engelson, "How to Get More Out of Your Spectrum Analyzer," *Microwaves,* May 1963.

Garrett, "Millimeter-Wave Spectrum Analyzer," *Microwave System Designer's Handbook,* June 1984.

Kicheloe, "The Measurement of Frequency with Scanning Spectrum Analyzers," Stanford Electronics Laboratories Report no. SEC-62-098, 1962.

Montgomery, "Technique of Microwave Measurements," *Radiation Laboratory Series,* Vol. XI. New York: McGraw-Hill or Boston Technical Publishers, 1964.

Myer, "Getting Acquainted with Spectrum Analyzers," Tektronix no. A-2273-2.

"Spectrum Analyzer Basics," Hewlett-Packard no. AN 150.

"The Tektronix Spectrum Analysis Primer," Tektronix no. 26W-4515-2.

Modulation

Black, *Modulating Theory,* Van Nostrand, 1953.

Charest, "Measuring Wide-Bandwidth FM Deviation," *EDN,* March 1, 1969.

Engelson and Breaker, "Interpreting Incidental FM Specifications," *Frequency Technology,* February 1969.

Giacoletto, "Generalized Theory of Multitone Amplitude and Frequency Modulation," *Proc. IRE,* July 1947.

Harvey, *et al.,* "The Component Theory of Calculating Radio Spectra with Special Reference to FM," *Proc. IRE,* June 1951.

"Spectrum Analysis — AM and FM," Hewlett-Packard no. AN 150-1.

Random Noise

Blackman and Tukey, *The Measurement of Power Spectra.* New York: Dover Publications, 1958.

Davenport and Root, *An Introduction to the Theory of Random Signals and Noise.* New York: McGraw-Hill, 1958.

Engelson, "Noise Measurements Using the Spectrum Analyzer — Random Noise," Tektronix no. AX-3260.

"Measurement of White Noise Power Density," Hewlett-Packard Application Note 63C.

Peterson, "Response of Peak Voltmeters to Random Noise," *GR Experimenter,* vol. 31, no. 7, December 1956.

Sutcliffe, "Relative Merits of Quadratic and Linear Detectors in the Direct Measurement of Noise Spectra," *The Radio and Electronic Engineer,* February 1972.

"Spectrum Analysis — Noise Measurements," Hewlett-Packard App. Note 150-4.

"Useful Formulas, Tables, and Curves for Random Noise," *GR Instrument Notes* no. IN-103, 1967.

Frequency Stability

Ashley, et al., "The Measurement of Noise in Microwave Transmitters," IEEE, *PGMTT,* April 1977.

Barnes, "Characterization of Frequency Stability," *NBS Technical Note* 394, 1970.

Engelson, "Consider Stability for Better Spectrum Analysis," *Microwaves,* May 1978.

Engelson, I., "Pinning Down Frequency Stability," *EDN,* May 15, 1969.

Fisscher, "Analyze Noise Spectra with Tailored Test Gear," *Microwaves,* July 1979.

Gumm and Engelson, "Spectrum Analysis of the TV and FM Signal," *BM/E,* April 1973.

Hamilton, "FM and AM Noise in Microwave Oscillators," *Microwave Journal,* June 1978.

Howe, "Frequency Domain Stability Measurements: A Tutorial Introduction," NBS Technical Note 679, 1976.

Kauffman and Engelson, "Frequency Domain Stability Measurements," *The Microwave Journal,* May 1967.

Lance, et al., Automated Phase Noise Measurements," *Microwave Journal,* June 1977.

Leeson, "A Simple Model of Feedback Oscillator Noise Spectrum," *Proc. IEEE,* February 1966.

Manassewitch, "Frequency Synthesizers," Wiley-Interscience, 1976.

Owen, "Measurement of Phase Noise in Signal Generators," *Marconi Instrumentation,* vol. 15, no. 6.

Rutman, "Characterization of Phase and Frequency Instabilities in Precision Frequency Sources," *Proc. IEEE,* September 1978.

Scherer, "Learn About Low Noise Design," *Microwaves,* April/May 1979.

Shields, "Review of the Specifications and Measurement of Short-Term Stability," *Microwave Journal,* June 1969.

Shoaf, et al., "Frequency Stability Specification and Measurement: High Frequency and Microwave Signals," *NBS Technical Note* 632, January 1973.

Vessot, et al., "The Specification of Oscillator Characteristics from Measurements Made in the Frequency Domain," *Proc. IEEE*, February 1966.

Watkins-Johnson, "Local Oscillator Phase Noise," *Tech-Notes*, vol. 8, no. 6, November/December 1981.

"Understanding and Measuring Phase Noise in the Frequency Domain," Hewlett-Packard App. Note 207.

Pulses/Impulses

Andrews, "Impulse Generator Spectrum Amplitude Measurement Techniques," *IEEE Trans. Instrum. Meas.*, December 1976.

Andrews and Arthur, "Spectrum Amplitude, Definition, Generation and Measurement," *NBS* Technical Note 699, 1977.

Arthur, "Impulse Spectral Intensity — What Is It?," *NBS*, NBSIR 74-365, 1974.

Barnard, "EMI Measurements Using a Spectrum Analyzer," Tektronix no. 26W-4971.

Barnard, "Smart Spectrum Analyzer Boosts EMI Testing," *MSN*, April 1982.

Bridges, "Spectrum Analyzers in EMC Measurements," *EMC Technology*, April/June 1983.

Engelson, "Check Impulse Bandwidth by Trimming Pulse Rate," *Microwaves*, January 1979.

Engelson, "EMI Applications Using the Spectrum Analyzer," Tektronix no. AX-3406.

Engelson, "Interpreting Pulsed-RF Spectra," *Microwaves*, March 1969.

Engelson, "Noise Measurement Using the Spectrum Analyzer — Impulse Noise," Tektronix no. AX-3259.

Engelson and Breaker, "Spectrum Analysis of FM'ing Pulses," *Microwave Journal*, June 1969.

Garrett and Engelson, "Pulsed RF Spectrum Analysis," Tektronix no. AX-4217.

Larson, "Calibration of Radio Receivers to Measure Broadband Interference," *NBS*, NBSIR 73-335, 1973.

Metcalf, et al., "Investigation of Spectrum Signature Instrumentation," IEEE, *Trans. EMC-7*, No. 2, June 1965.

Palladino, "A New Method for the Spectral Density Calibration of Impulse Generators," IEEE, *Trans. EMC*, February 1971.

Sabaroff, "Impulse Excitation of a Cascade of Series Tuned Circuits," *Proc. IRE*, December 1944.

Sabaroff, "Impulse Spectrum Analysis for Calibration of Impulse Noise Generators," 1st Conf. of RFI Reduction, conducted by IIT, 1954.

Simpson, "Broadband Pulsed/CW Calibration Signal Standard for Field Intensity Meter Receivers," *NBS*, NBSIR 74-371, 1974.

Straus, "Testing Products Correctly Ensures EMI-SPEC Compliance," *EDN*, November 1981.

"IEEE Standard for Measurement of Impulse Strength and Impulse Bandwidth,' *IEEE*, Std. no. 376, 1975.

"Spectrum Analysis – Pulsed RF," Hewlett-Packard Appl Note 150-2.

Measurements

Balmforth, "Millimeter-Wave Spectrum Analysis," *Electronic Engineer*, December, 1982.

Brusch, "The Spectrum Analyzer and the Earth Station," *Broadcast Engineering*, March 1983.

Engelson, "Baseband Measurements Using the Spectrum Analyzer," Tektronix no. AX-3433.

Engelson, "Crystal Device Measurements Using the Spectrum Analyzer," Tektronix no. AX-3535.

Engelson, "Define Dynamic Range for Better Spectrum Analysis," *Microwaves*, August 1977.

Engelson and Garrett, "Digital Radio Measurements Using the Spectrum Analyzer," *Microwave Journal*, April 1980.

Engelson, "Making Measurements with the 492," *Tekscope*, vol. 12, no. 1, 1980.

Engelson, "Spectrum Analyzer Applications in Baseband Measurements," *Tekscope*, vol. 10, no. 3, 1978.

Engelson, "Swept Selective Level Measurements," Tektronix no. AX-3861.

Engelson, "The Spectrum Analyzer as a Frequency Selective Meter," Tektronix no. AX-3682.

Engelson, "The Tracking Generator/Spectrum Analyzer System," Tektronix no. AX-3281.

Engelson, "Understand Resolution for Better Spectrum Analysis," *Microwaves*, December 1974.

Gumm, "No Loose Ends – Revised," *The Tektronix Proof-of-Performance Program for CATV*, Tektronix no. 26W-4889.

Klipper, "Sensitivity of Crystal Video Receivers with RF Pre-Amplification," *Microwave Journal,* August 1965.

Mott *et al.*, "Automating Spectrum Analysis with the 7854," Tektronix no. 42AX-4862.

Rasmussen, "Video Frequency Broadcast Measurements," Tektronix no. AX-3323.

Smith, "Cable Transfer Impedance Testing and Performance," *EMC Technology,* vol. 2, no. 4, October/December 1983.

Schrock, "Troubleshooting Two-Way Radios with the Spectrum Analyzer," Tektronix no. AX-3842.

"Spectrum Analysis – Accuracy Improvement," Hewlett-Packard no. AN150-8.

"Spectrum Analysis – Distortion Measurements," Hewlett-Packard no. AN150-11.

"Spectrum Analysis – Field Strength," Hewlett-Packard no. AN150-10.

"Spectrum Analysis – Noise Figure Measurement," Hewlett-Packard no. AN150-9.

"Spectrum Analysis – Signal Enhancement," Hewlett-Packard no. AN150-7.

"Spectrum Analysis – Tracking Generators," Hewlett-Packard no. AN150-3.

Spectrum Analyzer Techniques Handbook, Polarad Electronics Corp., 1955.

"Spectrum Analysis," Hewlett-Packard App. Note 63.

"Time and Frequency Domain Measurements," Hewlett-Packard no. AN240-0.

"Using a Narrow Band Analyzer for Characterizing Audio Products," Hewlett-Packard no. AN192.

INDEX

AM/FM combined, 81, 143
AM/FM, multitone, 81
AM or FM, 131
Amplitude, instantaneous, 70
Amplitude loss factor, 93
Amplitude measurement error, 219
Amplitude modulation, 70, 71, 121
   basic relationships, 121
   double sideband, 121
   frequency domain, 121
   multitone, 123
   single sideband, 124
   suppressed carrier, 123
Amplitude spectral density, 149
Angle modulation, 73
Angular frequency, 17
Angular velocity, 17, 19
Appendix, 225
Area of impulse, 55

Bandwidth,
   impulse, 102, 153, 173, 214
   random noise, 186, 214
   relationships, 93
   resolution, 233
Bel, 226
Bessel functions, 14, 16, 76, 129, 228
   small argument, 78
Bessel null, 135
   changing amplitude, 138
   changing frequency, 136
Boltzman's constant, 99

Carrier, 69
   null, 79
Circular functions, 17
Circuit response, 25
Combined amplitude, 34
Complex conjugate, 28
Complex notation, 20, 28
Complex plane, 20
Component transfer characteristic, 193
Contiguous filters, 26
Continuous spectrum, 8
Control setting, effect of, 164
   dense & line spectra, 164
   desensitization factor, 166
   dynamic range, 166
   fine detail, 166
   intensification effects, 168
   sensitivity, 166
   shape facor, 170
   skirt selectivity, 170
   transient response ringing, 169
Convolution, 89, 102
   graphical illustration, 104
Cyclical motion, 11

dB, 225
dBm, 225
DC pulse, 149
Decibel, 226
Degree of modulation, AM, 70
Delta function, 36

Dense spectrum, 8, 96, 149
   resolution setting, 96
Deviation, 75
Deviation linearity, 138
Digital radio signals, 204
Dirac function, 36, 55
Dirichlet conditions, 23
Discrete components, 8
Discrete pulse spectrum, 149
Dispersion, 2
Display flatness, 219
Distortion, 192
Drift, 214
Dynamic range, 220
   display, 220
   measurement, 220
   intermodulation, 220
   optimum, 221

EMI, 197
Energy distribution, 96
Euler's identity, 28
Exponential function, 225

Fast Fourier Transform, 2
FCC mask, 205
Filter bank, contiguous, 2
FM, *see* frequency modulation
Force-free solution, 25
Fourier analysis, 33, 52, 231
Fourier coefficients, 28, 33, 52
Fourier examples, 57
Fourier expansion, 23
Fourier integral, 49
Fourier transform pair, 49
Fourier relationships, 58, 59
Fourier series, 22, 33
   applications, 35
   arbitrary period, 34
   complex notation, 35
   least squared error, 24

spectral line existence, 24
   truncated, 23, 28
Frequency, 11, 17, 19
Frequency deviation, 75
Frequency domain, 1
   illustration, 25
Frequency identification, 113
Frequency instantaneous, 73
Frequency modulation, 74, 129
   argument, 129
   basic relationships, 129
   carrier null, 135
   effect on pulses, 157
   first sideband null, 141
   intensification effects, 146
   modulation index, 129
   multitone, 145
   narrowband, 131
   order, 129
   spectrum, 78
   ultrawideband, 141
   unresolved spectrum, 142
   wideband, 135
Frequency, negative, 51
Frequency response, 219
Frequency span, 2
Frequency translation, 4, 89, 102
Fundamental term, 23

Gaussian response, 93, 95
Gibb's phenomenon, 42

Harmonic amplitudes, 52
Harmonic conversion, 110
Harmonic motion, 17
Harmonic terms, 23
Harmonic zeros, 45, 52
Harmonics, infinite, 47

IEC specification, 209
IEEE specification, 209

Index

IF feedthrough, 109, 112
Image, 9, 110, 112
Imaginary numbers, 20
Impulse,
  approximation, 56
  area, 37
  bandwidth, see bandwidth, impulse,
  Fourier transform, 55
  frequency, 56
  function, 36, 54
  energy, 55
  noise, 185
  pseudo, 91
  strength, 37, 55
  symbolic, 38
Incidental FM, 215
Initial phase, 21
Input noise, equivalent, 210
Intercept point, 220
Intermodulation, 126, 238

Line Spectrum, 8
Linear time-invariant circuit, 24, 25
Lobe, main, 63, 98
Long pulse, CW, 98

Measurement interval, 96
Measurement limitations, 108
Measurement problem, 107
Missing pulses, 160
Mixer, 5
Modulation theory, 69

Negative frequency, 21, 22
Network response, sum, 39
Noise figure, 209
  calculations, 203
Noise power, 99, 210
Noise vs. bandwidth, 99
Noise sidebands, phase, 217

On/off ratio, 161
Optimum dynamic range, 221
Optimum resolution, 95
Optimum signal level, 221
Orthogonal functions, 12, 27
  examples, 14
  geometric illustration, 15
  set, 52
Oscillatory, 11
Overshoot, 43

Pulse area, 48
Pulsed rf, 6, 149, 234
  sensitivity, 236
Pulsed signal, 37, 95
  sensitivity, 100
Pulse desensitization factor, 153
Pulses, 149
  shape, 152, 155
  measurement, 151
  rectangle, 63
  trapezoid, 155
  triangle, 156
  sine squared, 157
Pulse signal, time-frequency diagram, 97
Pulse spectrum dc/rf ratio, 172
Pulse spectrum, negative
  frequencies, 171
  real and imaginary parts, 171
Pulse theory and measurement, 171
Pulse train, 44, 46, 96
Pulse vs. CW, 96, 99
Pulsewidth, 91
  parameters, 93

Radar performance, 154
Radian frequency, 17
Random noise, 183, 237
  absolute measurement, 185
  bandwidth, 186
  detectors, 184, 186
    examples, 189

input limitations, 187
interval to SA, 187
measurement, 183, 187
S/N ratio, 187
vertical display law, 186
video filter, 186
Reciprocal spreading, 56
Repetition rate lines, 97, 152
Residual response, 115
Resolution, 210, 215
    bandwidth, 233
    effect of, 85
    optimum, 213, 234
    calculation, 105
    distortion, 93
    equal signals, 211
    unequal signals, 212
Response, CW, 90
Response of circuits, 44, 85

Sawtooth waves, sum, 41
Sensitivity, 99, 209
    CW, 231
    calculation, 103
    improvement, 203
    loss function, 99
Shape factor, 212
Sidebands, AM, 71
Sidebands, power line, 219
Signal types, 6
Sine integral, 62
Sinewave, 14
Sinewave generation, 17
Sinewave, infinite duration, 37
Sinewave properties, 16
Sin x/x, 38
Sinusoid, frequency domain representation, 22
Skirt resolution, 212
S + N = 2N, 210
Span, 2
Spectral density distribution, 48

Spectral lines, 26
Spectrum analyzers, 1
Spectrum continuous, 44
Spectrum theory, 11
Spurious responses, 9
Squarewave configurations, 13
Squarewave harmonics, 39, 180
Squarewave on carrier, 182
Squarewave symmetry, 180, 181
Stability, 214
Steady state response, 8, 25
Superheterodyne arrangements, 5
Superheterodyne equations, 6
Superposition, 19, 39
Suppressed carrier AM, 73
Sweeping signal, 5
Sweeping signal spectrum, 89
Swept IF spurious responses, 108
Swept front-end spurious responses, 111
Symbols, 239
Synchronized sweeper, 194

T/t_o, 45
Telemetry subcarriers, 198
    proportional bands table, 199
Theory and measurement, 50
Time domain, 1
Time-frequency diagram, 92, 97
Time position, 91
Time, Newton's absolute, 1
Time variable circuit, 24, 29
Tracking generator, 194
Transducers, 201
Transient response, 8, 25, 85
Trigonometric notation, 21
Tuned filter system, 4
Two-two (2x2) response, 116
Types of instruments, 107
Types of measurements, 107

Unit impulse, 54

Index

Vector, rotating, 19
Vestigial sideband, 73

Waveform analysis, 179

Zero span, use of, 143

DATE DUE

HETERICK MEMORIAL LIBRARY
621.380433 E57m 1984   onuu
Engelson, Morris/Modern spectrum analyze

3 5111 00135 6751